Written by Clive Gifford
Illustrated by David Cousens

D1381584

First published 2009 by Kingfisher
an imprint of Macmillan Children's Books
a division of Macmillan Publishers Limited
The Macmillan Building, 4 Crinan Street, London N1 9XW
Basingstoke and Oxford
Associated companies throughout the world
www.panmacmillan.com

Additional illustrations by Sarah Cousens

ISBN 978-0-7534-1737-9

Copyright © Macmillan Children's Books 2009

All rights reserved. No part of this publication may be reproduced, stored in or introduced into a retrieval system, or transmitted, in any form or by any means (electronic, mechanical, photocopying, recording or otherwise), without the prior written permission of the publisher. Any person who does any unauthorized act in relation to this publication may be liable to criminal prosecution and civil claims for damages.

9 8 7 6 5 4 3 2 1
1TR/0509/TWP/MA/150MA/C

A CIP catalogue record for this book is available from the British Library.

Printed in Singapore

This book is sold subject to the condition that it shall not, by way of trade or otherwise, be lent, resold, hired out, or otherwise circulated without the publisher's prior consent in any form of binding or cover other than that in which it is published and without a similar condition including this condition being imposed on the subsequent purchaser.

Marie Curie

Contents

From ancient to modern 4

Archimedes 6

Galileo Galilei 10

Benjamin Franklin 16

James Watt 22

Isambard Kingdom Brunel 26

Thomas Edison 32

Nikola Tesla 38

Marie Curie 44

Glenn Curtiss 50

Sergei Korolev 54

Other famous inventors 60

Glossary 62

Index 64

From ancient to modern

Humans have always been fascinated about the world around them and how to alter it to their advantage. They learned to shape and manipulate wood and stone, and later metals, to create objects that made their lives safer or easier. Before there were great civilizations, many weapons and vital tools – from the sword and the plough to the hammer and the wheel – had already been invented.

Powerful ancient civilizations in China, Europe and the Muslim world were responsible for many important inventions. From China came the first compass, rockets, gunpowder and the first 'seismometer', or earthquake detector. And about 2,000 years ago, a Chinese court official, Cai Lun, came up with the first efficient paper-making process. Muslim inventors developed the first fountain pen with its own ink reserve, early torpedoes and windmills, and the pinhole camera, believed to have been invented by the astronomer Ibn-al-Haitham (965–1039CE).

An invention starts with an idea, but it may take many years of hard work to turn it into a working version. James Dyson, for example, invented the bagless vacuum cleaner in the 1980s, after he had built a staggering 5,127 test versions. Sometimes rivals battle it out to be thought of as the first inventor or to make a profit from selling their new invention – and sometimes those with the best business skills win. In many cases, inventors build on the work of others. The cars, computers, televisions and aircraft we use today are built on the knowledge and inventions of dozens of people from the past.

Inventors continue to alter the way we live our lives, but most are now based in the giant research centres of companies and national organizations. Occasionally, though, one man or woman's brilliance can create something of great everyday use – people such as Laszlo Biro, the inventor of the first reliable ballpoint pen, Mary Anderson who invented the car windscreen wiper or Trevor Baylis and his wind-up radio, which requires no batteries.

Archimedes

King Hiero II of Syracuse suspected theft. He had given gold to a craftsman to make a crown, but guessed that the man had kept some of it for himself, mixing the gold for the crown with less valuable metals such as brass or silver. But could the king prove it? He turned to the most gifted thinker in his kingdom, Archimedes.

Archimedes submerged the crown in a pot full of water. He knew that the volume of water pushed out (displaced) by the crown was equal to the crown's volume. He collected and measured the displaced water and then he weighed the crown. Using the measurements for the crown's volume and weight, Archimedes was able to calculate the crown's density. By the same method, he worked out the density of a lump of real gold of the same weight as the crown. The crown was less dense than the lump of gold – because it wasn't pure gold. The goldsmith had indeed cheated the king!

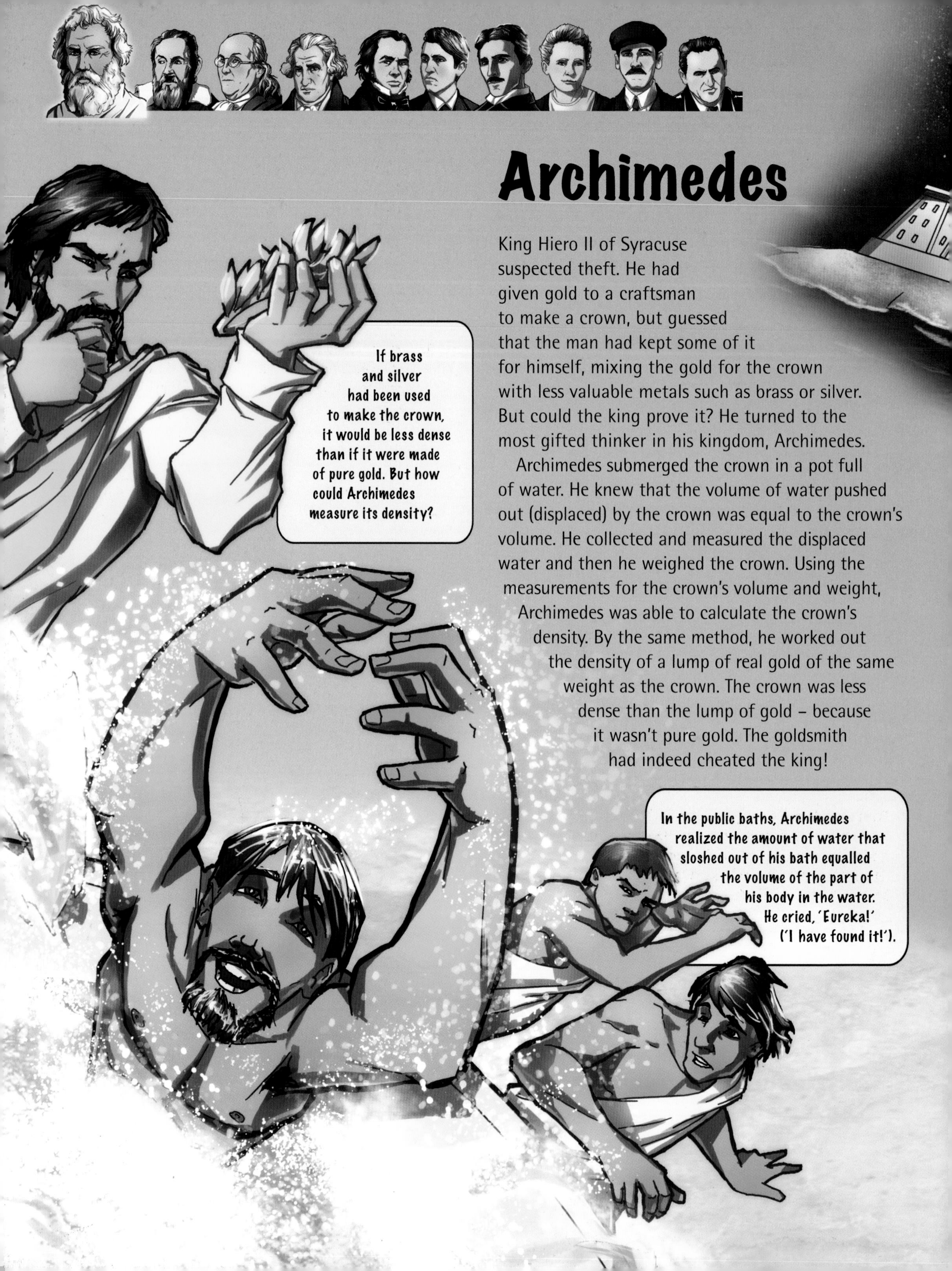

An Archimedes screw is a large screw inside a cylinder. When its handle is turned, water is drawn up the thread of the screw, raising it from the river onto the bank.

Archimedes was born in Syracuse on the island of Sicily. In his teenage years, he studied at Alexandria, Egypt, home to the world's biggest library. There, he read hungrily the works of great mathematicians such as Euclid, and listened avidly to lectures by his teacher and friend, Conon of Samos. It was in Egypt that Archimedes may have created his first major invention – the Archimedes screw – while watching workers struggle to lift water out of the river Nile.

Archimedes returned to Syracuse where he would live the rest of his life, devoting himself to mathematics and bursts of invention.

King Hiero II asked him to design one of the biggest ships in the ancient world. The *Syracusia* was said to be 55m long and could carry 600 people. But the ship was almost too heavy to launch. Archimedes had a solution. He had been experimenting with levers and pulleys, which enabled a few men to lift great weights. When Archimedes boasted about the power of his levers, Hiero asked him to prove it by launching a ship from his fleet.

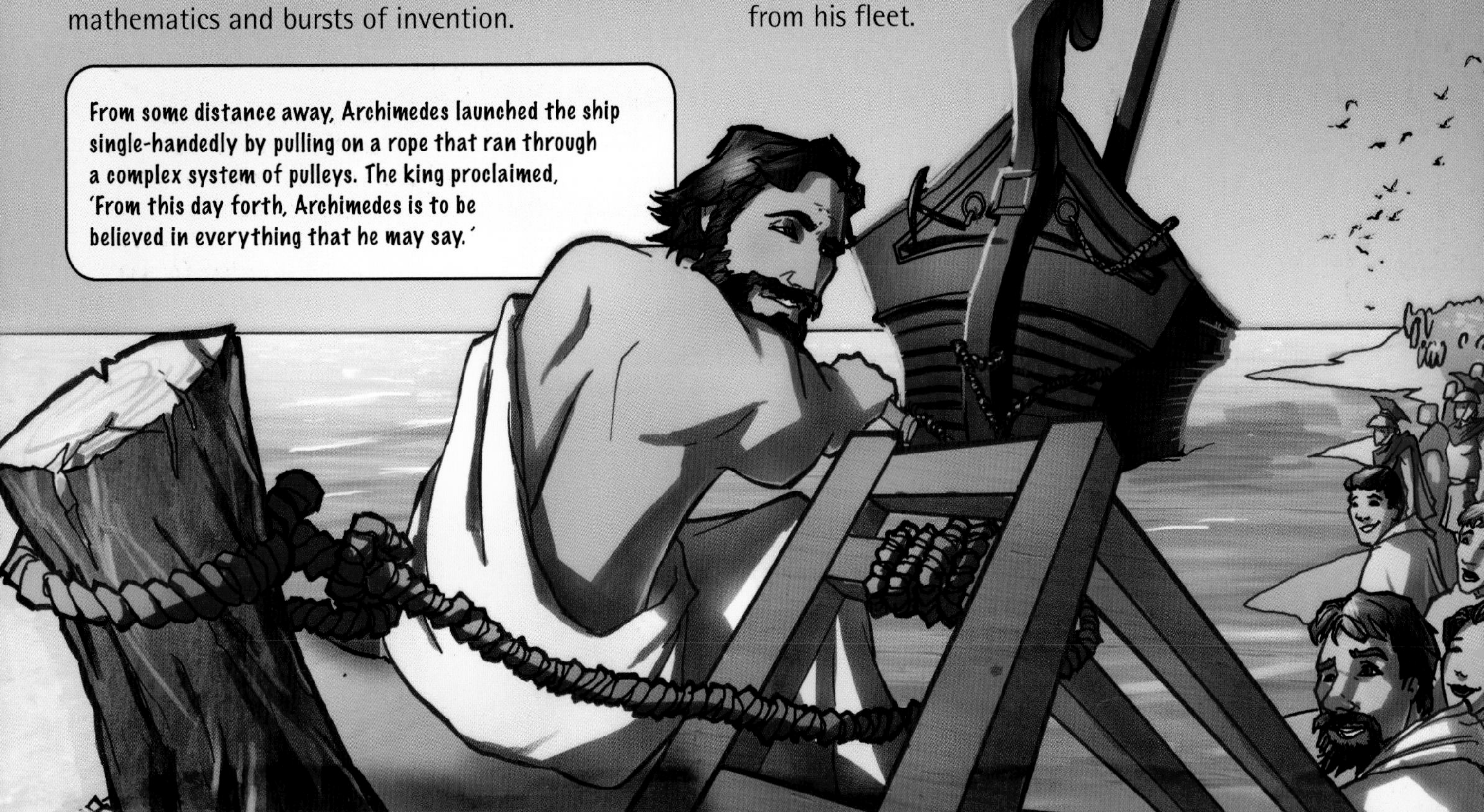

As a mathematician, Archimedes was more advanced than any other scholar of his era. He wrote about all aspects of mathematics, especially geometry (the mathematical study of shapes), making many breakthroughs. He worked out how to measure circles using formulae and established an accurate value for pi (π). His brilliant work *On the Sphere and the Cylinder* showed how to measure the volume and surface area of a sphere (a perfect ball shape). He also explained a sphere's relationship to a cylinder of the same height and diameter.

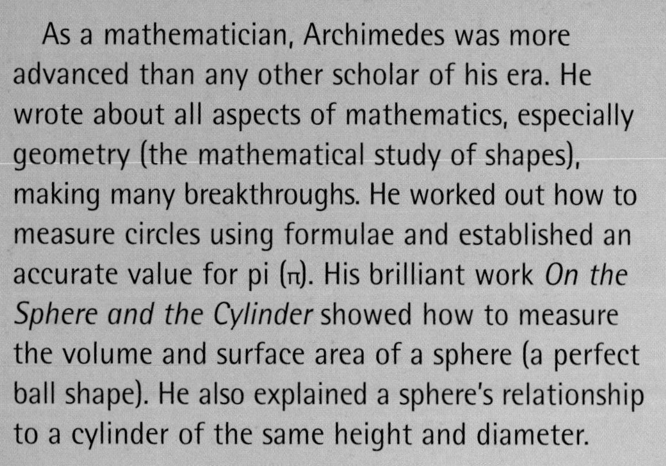

Archimedes worked for days without baths, food – even water. Servants would wash him as he continued to write theories and draw diagrams.

About 60 Roman ships attacked the city of Syracuse, but they were no match for the giant cranes that Archimedes had invented. Iron 'claws', lowered from the city's walls, hooked onto the ships. When chains were dragged through pulleys, the hooks were raised, toppling the ships into the sea.

King Hiero II died in 215BCE and under his successor, Hieronymos, Syracuse was bombarded by a Roman force led by Marcus Claudius Marcellus. With about 15,000 soldiers and 60 war galleys, Marcellus attacked from both sea and land. But he was to receive a nasty shock – for Archimedes, despite being in his seventies, had dramatically reinforced the city's defences. His catapults fired rocks and iron darts at the Roman ships, but most terrifying of all was the Claw of Archimedes – a giant crane that toppled ships as they neared the city's walls.

Archimedes' machines played a major part in repelling the Romans. Marcellus was forced to call off the siege, but in 212BCE he managed to seize control of the city. As the Roman soldiers sacked Syracuse, Marcellus ordered his troops to 'spare the mathematician'. He was ignored. One of the greatest inventors and mathematical thinkers of the ancient world was killed, slain by a Roman sword.

LIFE LINK

Galileo studied the works of Archimedes avidly. He invented a method of raising water, as Archimedes had done, and his book *La Balancitta* (*The Little Balance*) described Archimedes' methods of studying the densities of objects with a balance. He called the ancient Greek 'the superhuman Archimedes whose name I never mention without a feeling of awe'.

Accounts of Archimedes' death differ. One states that he was killed by a Roman soldier because he wouldn't stop drawing diagrams in the sand. Marcellus mourned for Archimedes and had him buried with full honours.

Galileo Galilei

Rome, 1633. Galileo was in prison, on trial by the dreaded Inquisition. Week after week of intense questioning had left him shattered, his nerves on a knife edge. The Inquisition investigated those it believed were against the Catholic Church. Many of its victims were tortured in horrific ways. Galileo Galilei had escaped that fate... so far.

Galileo was born in the Italian city of Pisa in 1564, the first of seven children of Vincenzio Galilei. As a boy, Galileo studied Greek and drawing as well as music. At the age of 15, after four years of schooling at a monastery, he declared he wanted to be a monk. His father refused, trying to force him into the clothing trade, but eventually, at 17, Galileo was enrolled as a medical student at the University of Pisa.

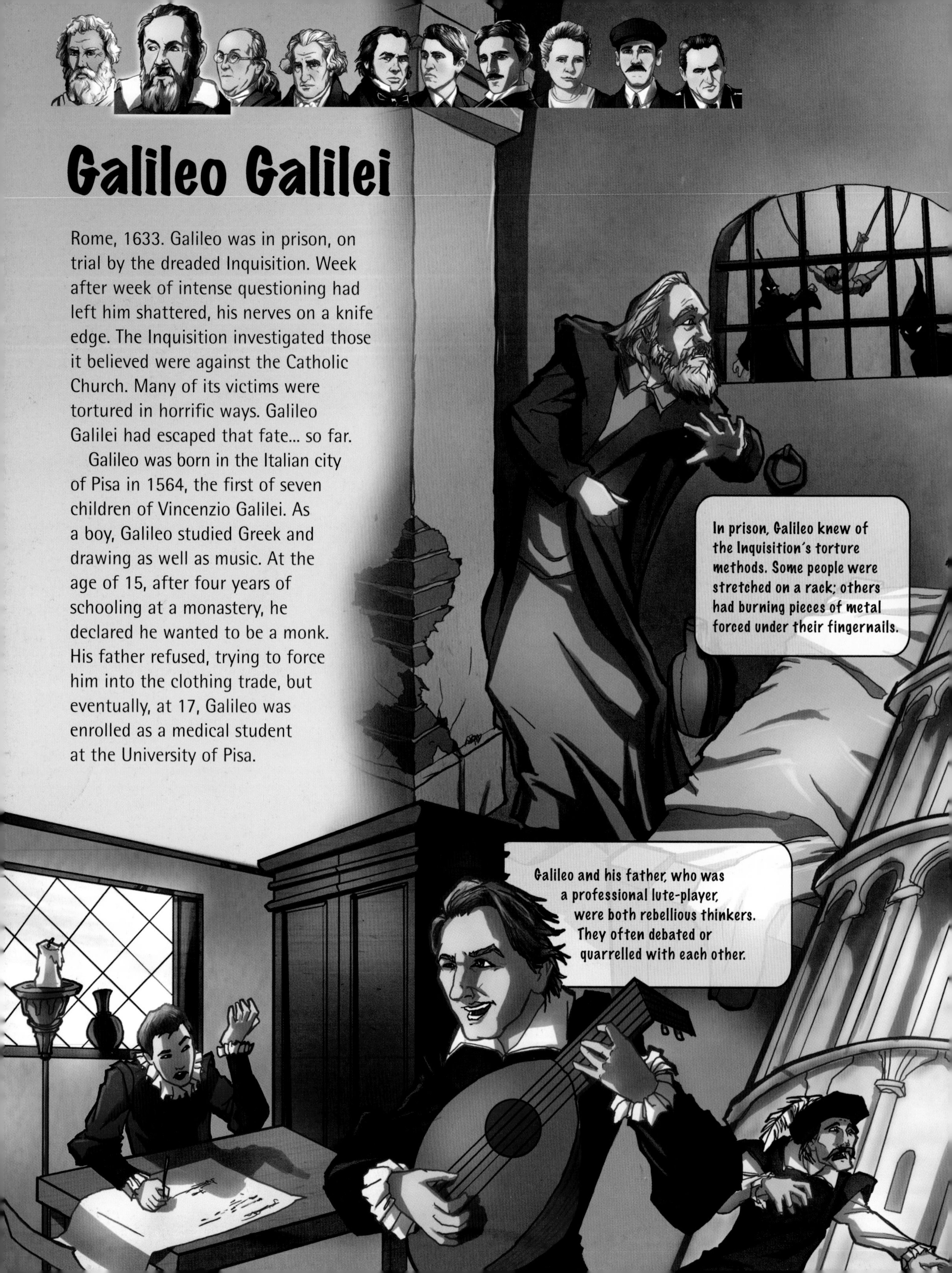

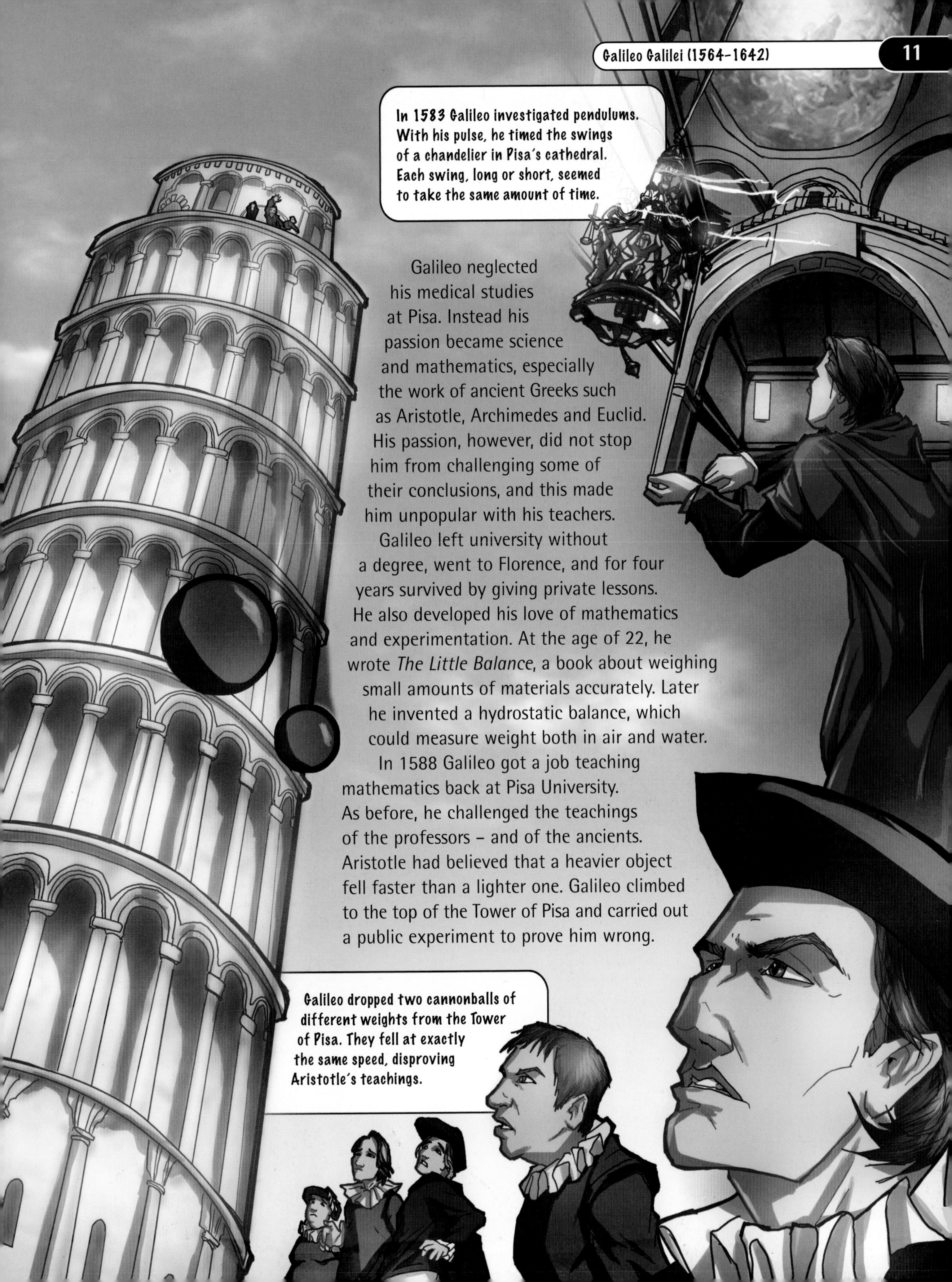

Galileo neglected his medical studies at Pisa. Instead his passion became science and mathematics, especially the work of ancient Greeks such as Aristotle, Archimedes and Euclid. His passion, however, did not stop him from challenging some of their conclusions, and this made him unpopular with his teachers.

Galileo left university without a degree, went to Florence, and for four years survived by giving private lessons. He also developed his love of mathematics and experimentation. At the age of 22, he wrote *The Little Balance*, a book about weighing small amounts of materials accurately. Later he invented a hydrostatic balance, which could measure weight both in air and water.

In 1588 Galileo got a job teaching mathematics back at Pisa University. As before, he challenged the teachings of the professors – and of the ancients. Aristotle had believed that a heavier object fell faster than a lighter one. Galileo climbed to the top of the Tower of Pisa and carried out a public experiment to prove him wrong.

Galileo's questioning nature and his new scientific theories brought him into so much conflict in Pisa that he was forced to leave in 1592. He moved to Padua to become Professor of Mathematics at its university. There, people were more accepting of his ideas and Galileo settled happily for 18 years. Soon after arriving, he invented a pump for raising water and later he developed an instrument called a sector. In Padua, Galileo met Marina Gambina and, although they never married, they had three children, Virginia, Livia and Vincenzio. Galileo's salary also increased with his new job and this helped to fund his experiments. Every penny was needed as his father had died a few years before and Galileo was now head of the family.

Galileo's life changed in 1609. Word of a new invention from the Netherlands reached Padua – it was a telescope. Galileo, who may have seen an early version of it or worked out its principles by himself, started making lenses and building his own telescopes. His first one magnified objects three times while the one he presented to the Senate, who ran the state of Venice, offered nine-times magnification.

Galileo's telescopes were more powerful than the early types invented in the Netherlands and it was Galileo who began to use the instrument as a scientific tool. In the winter of 1609, he aimed his telescopes skyward. He was astonished by what he saw – craters on the Moon! Then in January 1610, Galileo observed Venus during the different phases of what was thought to be its journey around Earth. And he was stunned to discover four moons orbiting Jupiter. He wrote about these and other earth-shattering discoveries in a small but bestselling book, *The Starry Messenger*, and gave many lectures.

The Moon was thought to be smooth, but Galileo saw craters and mountains with his telescope. He sketched them for his book *The Starry Messenger*.

Later in 1610, Galileo moved to Florence to become the court mathematician for the powerful Medici family. As his fame spread throughout Italy, Galileo continued working, discovering spots on the Sun and rings around Saturn. Questions flooded his mind. The view of the Universe agreed by the Church was the one offered by Aristotle almost 2,000 years earlier – that everything revolved around Earth. A Polish astronomer, Nicolas Copernicus, had challenged this view in the 1500s, and Galileo's work supported Copernicus's view that the Earth moves around the Sun.

Galileo gave many lectures and demonstrations of his theories that Earth orbited the Sun and not the other way around. These displays did not go unnoticed, especially among church officials, and he made many enemies.

Galileo thought he could persuade the Catholic Church to abandon its long-held views. But its head, Pope Paul V, did not budge, even ordering an investigaton into Galileo. Then in 1616, it was declared that Earth was the centre of the Universe and that to believe otherwise was a serious crime against the Church. Galileo was forced to lie low. But in 1624, with a new pope in power, Galileo was able to visit Rome to discuss his beliefs openly. Pope Urban VIII agreed that Galileo could write a book about the Copernican system, but the book must devote equal importance to the view that the Universe orbited Earth. However, Galileo's *Dialogue Concerning the Two Chief Systems of the World* was far from balanced and the pope was furious. Galileo was ordered back to Rome in 1633 to stand trial for his beliefs. He was found guilty and sentenced to life imprisonment. Later he was allowed to return to his villa near Florence, but under house arrest. He would remain a virtual prisoner for the rest of his life.

When Galileo met with Pope Urban VIII, he was hopeful that his ideas would be accepted by this new leader of the Church, who had once written poetry praising 'learned Galileo'.

Galileo defended his theories before the Inquisition three times in June 1633. An elderly man of 69, throughout his trial he complained of 'poor health' and 'arthritic pains'.

Galileo was banned from publishing any scientific work, but eventually he defied the Church and returned to work. With his eyesight failing, he produced an influential book called *Two New Sciences*. His manuscript had to be smuggled by friends out of his house and out of Italy so that it could be published in the Netherlands. The book examined how objects moved, the forces that worked on them and how materials can be stretched and shaped. This work was as revolutionary as his astronomy and paved the way for a rich era in physics.

Galileo went completely blind in 1638 and relied on assistants to look after him. His son, Vincenzio, joined him for the last year of his life. Confined to bed with a fever and a kidney complaint, Galileo died on January 9th 1642.

Vincenzio helped his father return to the ideas about pendulums that he had developed as a youth. Together they investigated how a pendulum could be used to make a clock.

Confined to his villa, Galileo fell into a dark depression, deepened by the news of his daughter's death. For a long time, he ignored the scientific equipment that the Church had allowed him to keep.

LIFE LINK

Galileo's work with lenses enabled him to make telescopes powerful enough to revolutionize astronomy and our view of the Universe. In the early 1760s, Benjamin Franklin also worked with lenses. By joining together lenses of different strengths, he created bifocal spectacles.

Benjamin Franklin

As ten-year-old Ben Franklin skimmed the scum and dirt from the boiling vat of fat in his father's workshop, he dreamed of something more exciting than making candles and soap. One day, he would be one of America's greatest heroes, a scientist, an inventor and a statesman who would help draft the Declaration of Independence, announcing America's wish to be free from British rule. But for now, he was still under the firm control of his father.

Benjamin was the 15th child and youngest son of Josiah Franklin, who had sailed from England to the Americas in 1683. The Franklins were hardworking but poor, which made life in the crowded house tough. Ben thought of running away to sea. He was an excellent swimmer and had already made his first invention, a pair of flipper-like boards for his hands to speed up his swimming. Instead, his father apprenticed him, aged 12, to Benjamin's own brother, James, a printer. Franklin mastered printing skills rapidly and read the printed books whenever he could. But working for James was far from easy. James was envious of his younger brother's intelligence and often beat him.

At the age of 17, Ben fled to New York, then to Philadelphia, where he worked as a printer's assistant. Over the years, he built up his own printing business, opening a stationery shop in 1728 and taking over a newspaper, *The Pennsylvania Gazette*, the following year. But Franklin was more than just a businessman. He cared about his community passionately and, with eleven of his friends, he started the Junto, a club that met once a week to discuss ways to improve Philadelphia.

In 1736 Franklin and four friends set up a fire brigade of volunteers for the city. And a few years later, still mindful of the risks of house fires, Franklin invented a safer form of fireplace. Like all his inventions, he chose not to patent it, which would legally declare that the invention belonged to him and would make him a fortune. In his autobiography, Franklin explained, 'As we enjoy great advantages from the inventions of others, we should be glad of an opportunity to serve others by any invention of ours; and this we should do freely and generously.'

Franklin secretly wrote articles for his brother's newspaper, *The New England Courant*, signing them with fake names such as Alice Addertongue, Harry Meanwell and Mrs Silence Dogood.

The 'Franklin Stove' fireplace generated more heat and less smoke – and it was less likely to cause house fires.

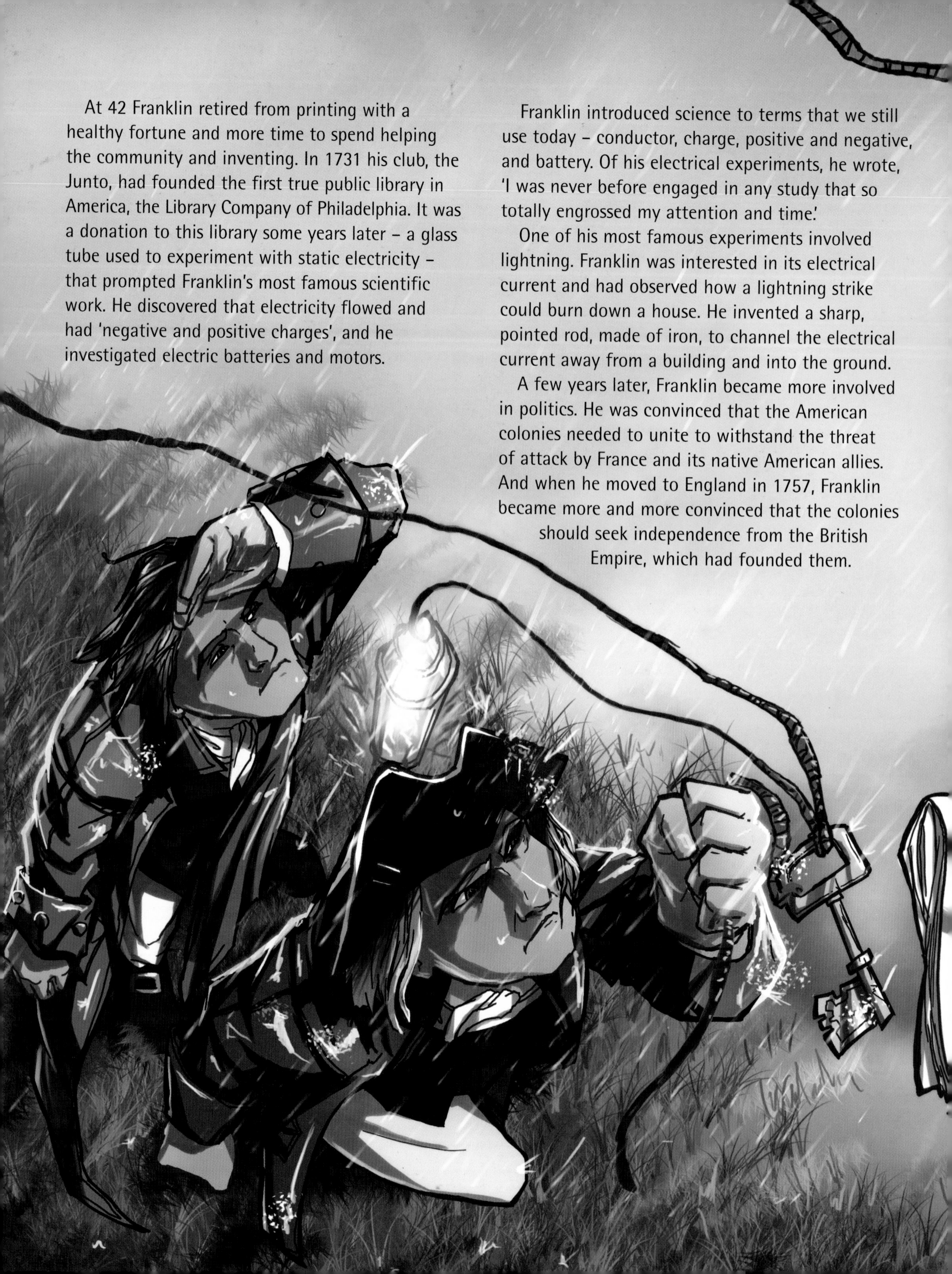

At 42 Franklin retired from printing with a healthy fortune and more time to spend helping the community and inventing. In 1731 his club, the Junto, had founded the first true public library in America, the Library Company of Philadelphia. It was a donation to this library some years later – a glass tube used to experiment with static electricity – that prompted Franklin's most famous scientific work. He discovered that electricity flowed and had 'negative and positive charges', and he investigated electric batteries and motors.

Franklin introduced science to terms that we still use today – conductor, charge, positive and negative, and battery. Of his electrical experiments, he wrote, 'I was never before engaged in any study that so totally engrossed my attention and time.'

One of his most famous experiments involved lightning. Franklin was interested in its electrical current and had observed how a lightning strike could burn down a house. He invented a sharp, pointed rod, made of iron, to channel the electrical current away from a building and into the ground.

A few years later, Franklin became more involved in politics. He was convinced that the American colonies needed to unite to withstand the threat of attack by France and its native American allies. And when he moved to England in 1757, Franklin became more and more convinced that the colonies should seek independence from the British Empire, which had founded them.

Franklin continued to invent while he was in England. He declared that his armonica, a musical instrument, gave him the most pleasure of all his inventions. It had glass bowls that could be played with wet fingers like the rims of wine glasses. The bowls produced notes that Franklin described as 'incomparably sweet'.

Franklin and his 21-year-old son William probably carried out their legendary lightning experiment in 1752. They made a kite from a silk handkerchief and fitted it with a pointed wire, which attracted lightning. The electric charge travelled down some twine to a metal key, creating sparks of electricity. Franklin stored the electricity in a device called a Leyden jar.

Published in *The Pennsylvania Gazette* in 1754, Franklin's 'Join or Die' sketch was the first political cartoon in America. The snake divided into parts represented the different American colonies, which, he felt, must join together to survive.

Franklin premiered his armonica in 1762 when Marianne Davies played it at a concert in London. Both Mozart and Beethoven later composed music for the instrument.

Franklin's curiosity about the world and how to harness it to make people's lives better never ceased. He proposed Daylight Saving Time, mapped storms and charted the Gulf Stream (a strong current in the Atlantic Ocean), helping ships to sail from America to Europe more quickly. And when Franklin found he needed different spectacles for seeing long distance and for reading, he invented bifocal glasses, which contained half a lens for each purpose.

Back in America, Franklin became a highly respected politician. One of his greatest achievements was to reform the postal service. He even invented a distance-measuring device called an odometer, which enabled him to plot precise distances between post offices and to rearrange deliveries to be more efficient.

In 1776 Franklin sailed the Atlantic again, this time bound for France, where, as commissioner (ambassador) for the United States, he would stay for eight years. Many in French society were entranced by his wit, energy and brains. Franklin, in turn, enjoyed the company of France's great writers and scientists, and managed to seal crucial French support for America during their fight to be free of British rule. In 1783 Franklin signed the Treaty of Paris, which formally ended the American Revolution, recognizing the United States as an independent nation.

While in Paris in 1783, Franklin witnessed one of the first passenger-carrying balloon flights.

Franklin built a special chair for his beloved library. Its seat was reversible so it could also be used as a small step ladder.

Returning to America in 1785 as a hero, Franklin was showered with honours and letters of congratulation. Now almost 80, he remained active, involved with the creation of the United States Constitution (system of government) and speaking out against slavery. But he was also content to spend time at home in Philadelphia in his huge library, where he invented a mechanical arm to help him reach books on the top shelves. Franklin referred to himself as a printer despite his many achievements as a diplomat, statesman, writer, publisher and inventor. A modest man, he said, 'The noblest question in the world is what good may I do in it?'

LIFE LINK

Franklin met James Watt's business partner, Matthew Boulton, in England in 1759. He became a member of the Lunar Society, a group of gifted British scientists that included Joseph Priestley, Boulton and Watt. When Franklin learned of Watt's new copying machine, he ordered three of them, giving one to Thomas Jefferson.

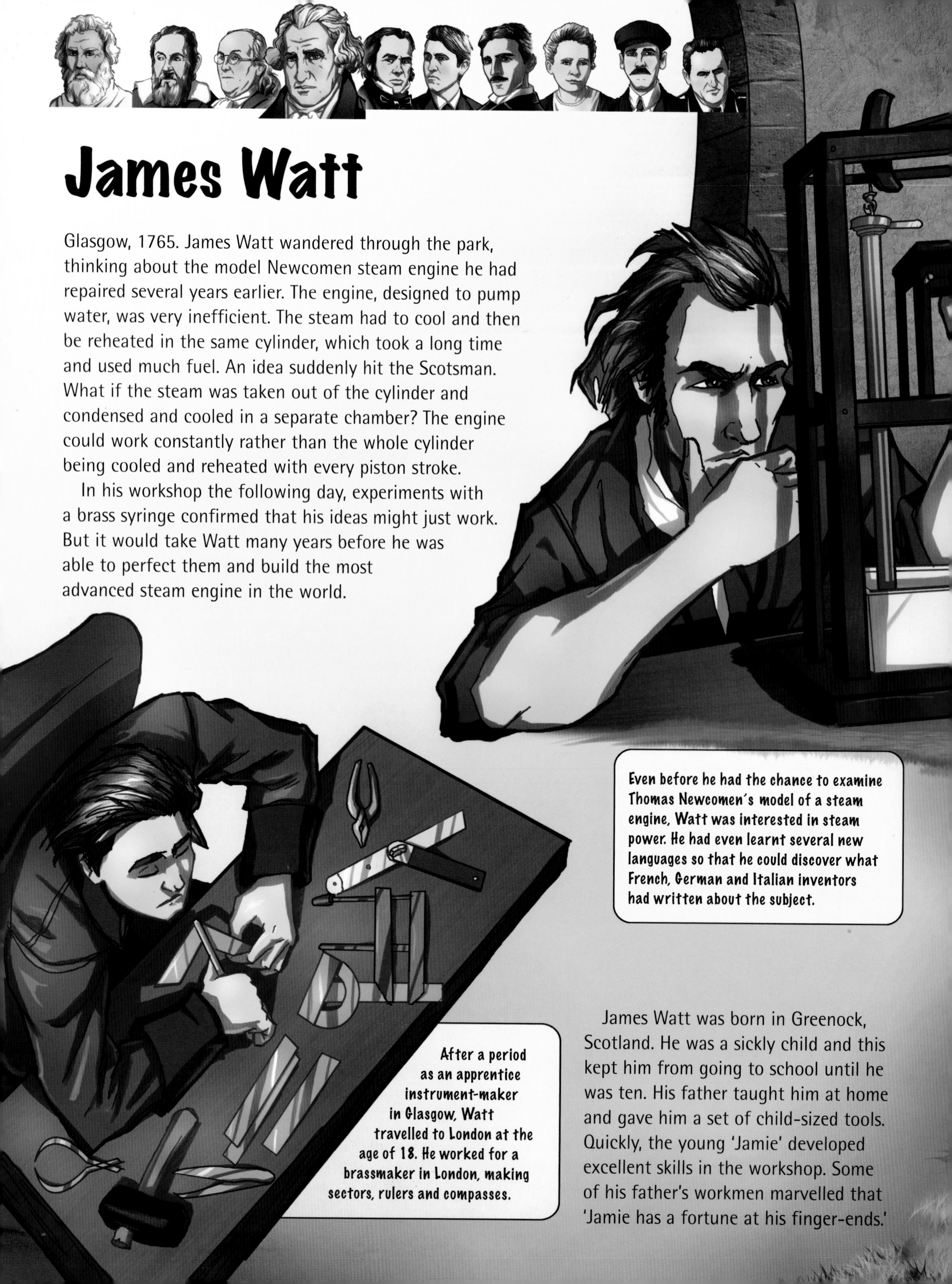

James Watt

Glasgow, 1765. James Watt wandered through the park, thinking about the model Newcomen steam engine he had repaired several years earlier. The engine, designed to pump water, was very inefficient. The steam had to cool and then be reheated in the same cylinder, which took a long time and used much fuel. An idea suddenly hit the Scotsman. What if the steam was taken out of the cylinder and condensed and cooled in a separate chamber? The engine could work constantly rather than the whole cylinder being cooled and reheated with every piston stroke.

In his workshop the following day, experiments with a brass syringe confirmed that his ideas might just work. But it would take Watt many years before he was able to perfect them and build the most advanced steam engine in the world.

Even before he had the chance to examine Thomas Newcomen's model of a steam engine, Watt was interested in steam power. He had even learnt several new languages so that he could discover what French, German and Italian inventors had written about the subject.

After a period as an apprentice instrument-maker in Glasgow, Watt travelled to London at the age of 18. He worked for a brassmaker in London, making sectors, rulers and compasses.

James Watt was born in Greenock, Scotland. He was a sickly child and this kept him from going to school until he was ten. His father taught him at home and gave him a set of child-sized tools. Quickly, the young 'Jamie' developed excellent skills in the workshop. Some of his father's workmen marvelled that 'Jamie has a fortune at his finger-ends.'

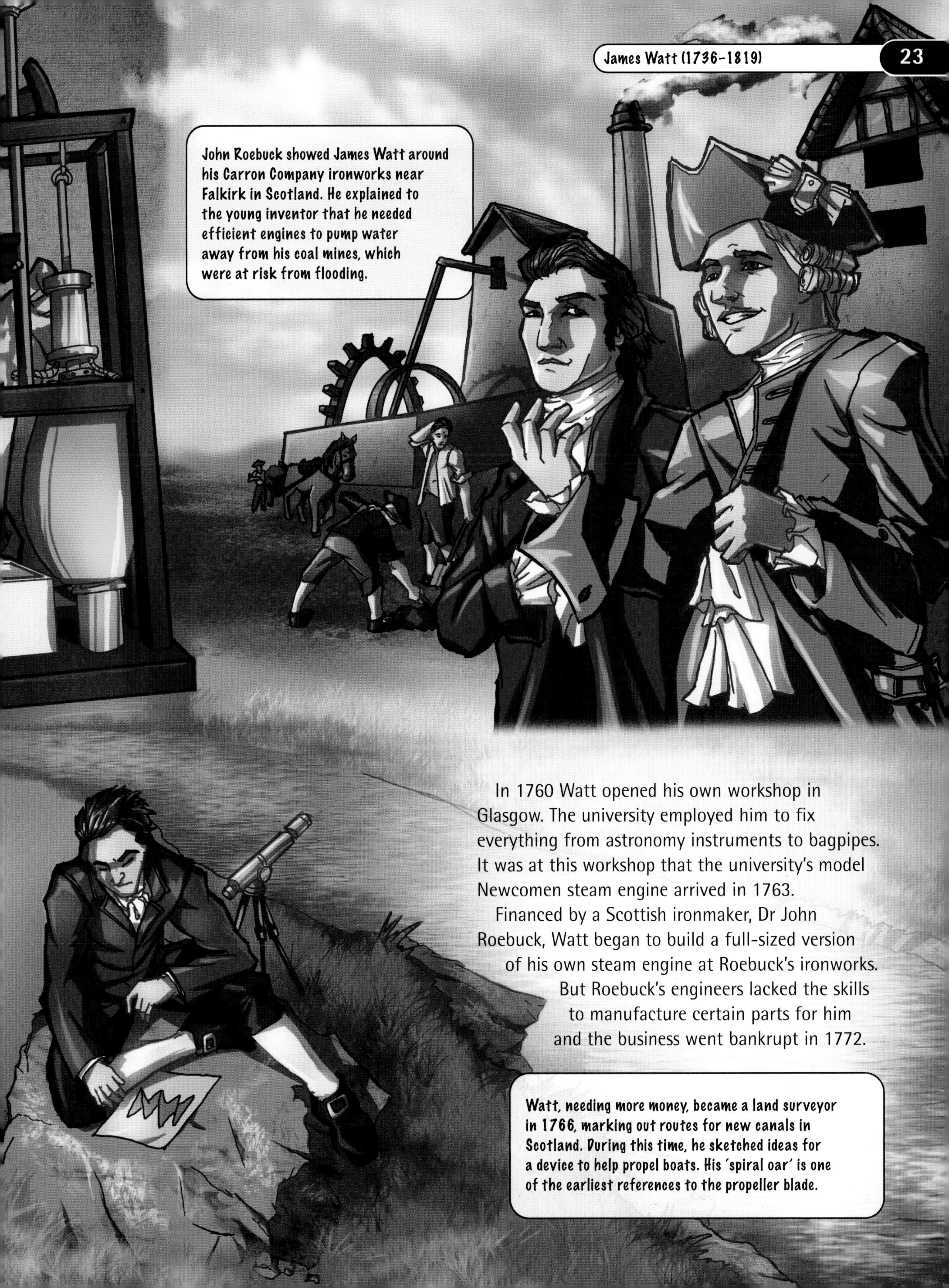

In 1760 Watt opened his own workshop in Glasgow. The university employed him to fix everything from astronomy instruments to bagpipes. It was at this workshop that the university's model Newcomen steam engine arrived in 1763.

Financed by a Scottish ironmaker, Dr John Roebuck, Watt began to build a full-sized version of his own steam engine at Roebuck's ironworks. But Roebuck's engineers lacked the skills to manufacture certain parts for him and the business went bankrupt in 1772.

In 1775 Watt became business partners with Matthew Boulton, an owner of a large Birmingham engineering works. The move would prove incredibly fruitful. In the same year, Watt secured patents for an improved steam engine, protecting his invention from being copied for 25 years. In fact Watt and Boulton fought so hard to prevent others from making similar devices, that they slowed industrial development in Britain. Boulton, a skilled engineer himself, urged Watt to develop an engine that could produce a circular movement for grinding, weaving and milling. Watt responded in 1781 by inventing the Sun-and-planet gearing system, which used cogwheels to help turn the up-and-down motion of pistons into a rotational movement. Later Watt invented a double-acting engine that generated power on both the up and down movements of a piston. He effectively doubled the power of his engines.

Watt retired in 1800, the same year that most of his patents ran out. By then many factories, including 84 cotton mills, used his steam engines. Now a wealthy man, he passed his share of the business to his son, James Watt Jr, travelled abroad with his wife and bought a country home in Wales. But he still tinkered with machinery in his workshop, trying to develop a 'copy mill' – a machine that could make copies of sculptures for his friends.

By the time of his death in 1819, Watt could see for himself the massive boom in steam-engine-powered machinery. It was his improvements to the steam engine that had helped usher in the industrial revolution. The world would not be the same again.

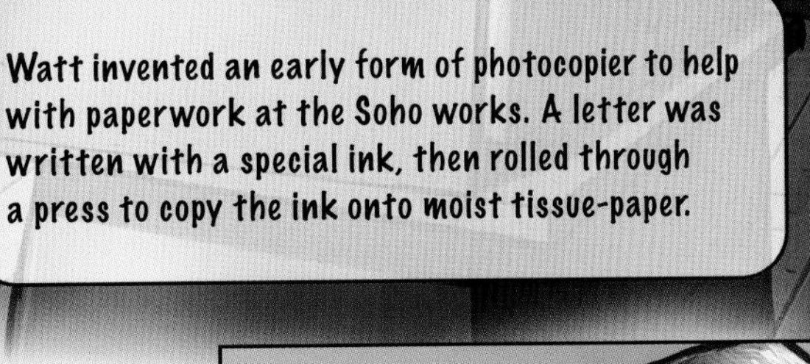

Watt invented an early form of photocopier to help with paperwork at the Soho works. A letter was written with a special ink, then rolled through a press to copy the ink onto moist tissue-paper.

In 1788 Watt developed the 'flyball governor' to keep engines moving at a constant speed. As the engine speeded up, two steel balls whizzed outwards, causing a collar to slide up the shaft and alter the engine speed.

Even away from his engines, Watt was an inquisitive man. He experimented with gases with famed scientist Joseph Priestley, a fellow member of the Lunar Society, a group of scientists, philosophers and intellectuals.

LIFE LINK

Watt's improvements to steam-engine technology hastened in an age of practical railway transport with railway lines built by Isambard Kingdom Brunel. Watt's son, James Watt Jr, assisted American efforts to build steamships, paving the way for Brunel's majestic ocean liners.

Isambard Kingdom Brunel

January 1827. Isambard Kingdom Brunel was working with two of his best miners, tunnelling deep below London's River Thames, when disaster struck. Water flooded in and a giant wave hurtled down the tunnel. Two years earlier, the young engineer had stood proudly alongside his father, Marc, as the pair laid the first bricks of the world's first tunnel to be built underneath a major river. Now the river had struck back. Brunel was knocked unconscious and swept along the tunnel's length...

Brunel's father was one of Europe's most talented engineers, and Brunel grew up learning about geometry and freehand drawing. At 14, he was sent to France to study at Caen College, a stronghold of mathematics. Then he was apprenticed in Paris to the famous designer of mechanical instruments, Louis Breguet.

Brunel was just 20 when his father appointed him Resident Engineer on the extraordinarily ambitious project to build a tunnel under the River Thames in London.

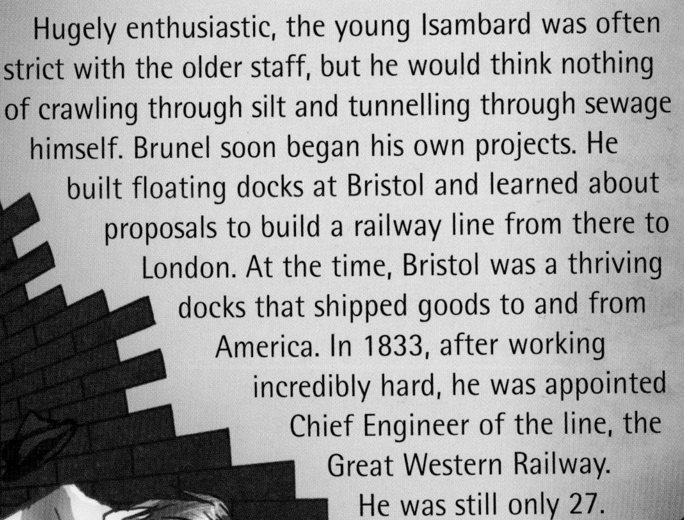

Hugely enthusiastic, the young Isambard was often strict with the older staff, but he would think nothing of crawling through silt and tunnelling through sewage himself. Brunel soon began his own projects. He built floating docks at Bristol and learned about proposals to build a railway line from there to London. At the time, Bristol was a thriving docks that shipped goods to and from America. In 1833, after working incredibly hard, he was appointed Chief Engineer of the line, the Great Western Railway. He was still only 27.

Isambard and Marc Brunel hosted the world's first underwater banquet in their Thames tunnel in 1827. The tunnel was finally completed in 1843. It is still used today and every year 14 million people travel through it on underground tube trains.

Brunel met Mary Horsley, his future wife, in 1832. It is said that he entertained her with displays of magic while the famous composer Felix Mendelssohn played the piano.

Brunel surveyed every kilometre of the Great Western Railway himself, racing on horseback between locations. He even designed his own mobile office – a long horse-drawn carriage with room for a bed and a drawing board.

Brunel threw himself into the massive task ahead. Opposition to the railway line – the longest in Britain – was strong, but Brunel visited every landowner on the route to ease their minds. He also gathered workers, equipment and materials, tried to raise money and sought approval from Parliament. This was a huge operation. The entire project was costed at 2.5 million pounds, an enormous sum at the time. In the end, like many of Brunel's projects, it cost almost three times as much.

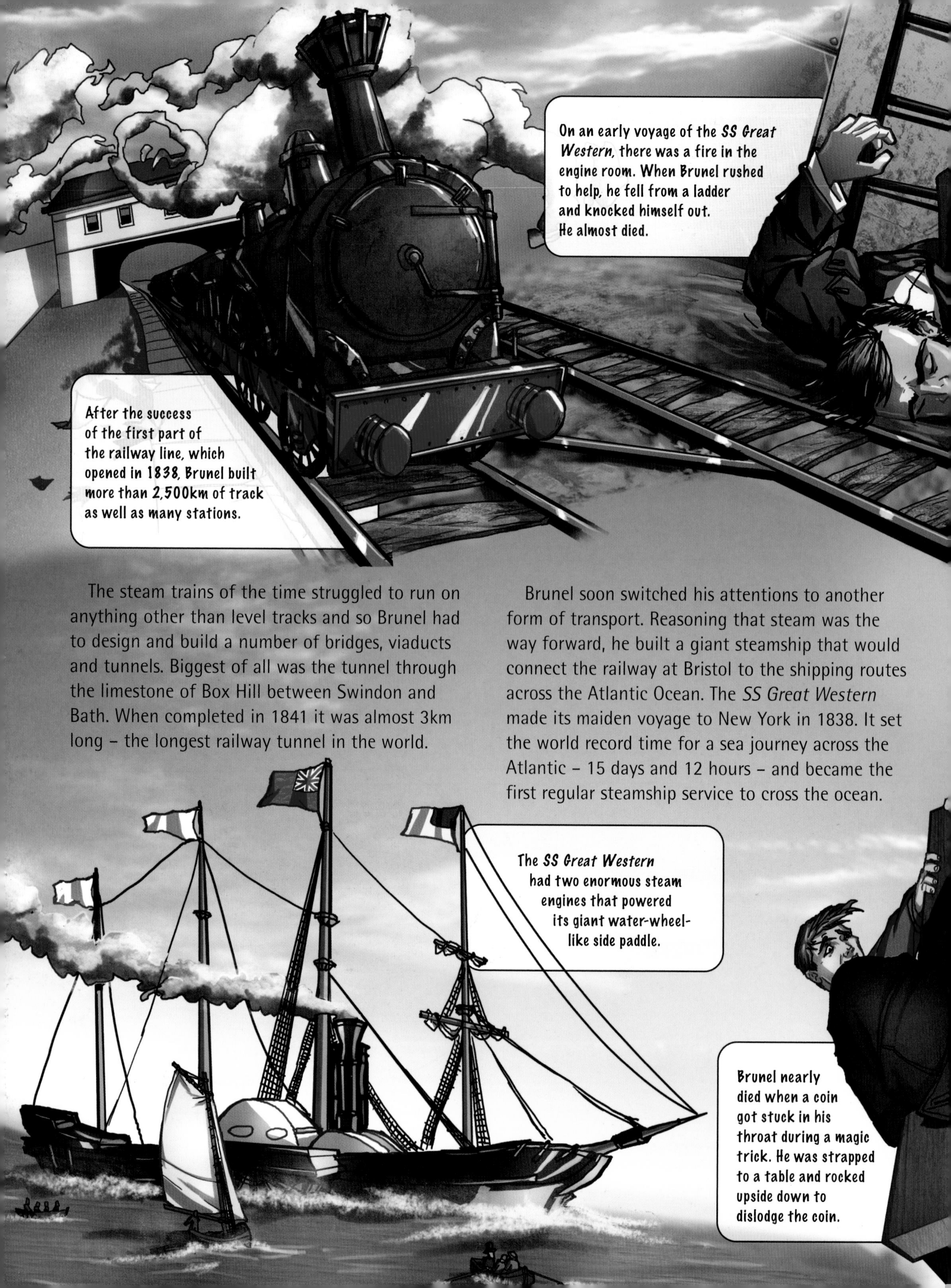

The steam trains of the time struggled to run on anything other than level tracks and so Brunel had to design and build a number of bridges, viaducts and tunnels. Biggest of all was the tunnel through the limestone of Box Hill between Swindon and Bath. When completed in 1841 it was almost 3km long – the longest railway tunnel in the world.

Brunel soon switched his attentions to another form of transport. Reasoning that steam was the way forward, he built a giant steamship that would connect the railway at Bristol to the shipping routes across the Atlantic Ocean. The *SS Great Western* made its maiden voyage to New York in 1838. It set the world record time for a sea journey across the Atlantic – 15 days and 12 hours – and became the first regular steamship service to cross the ocean.

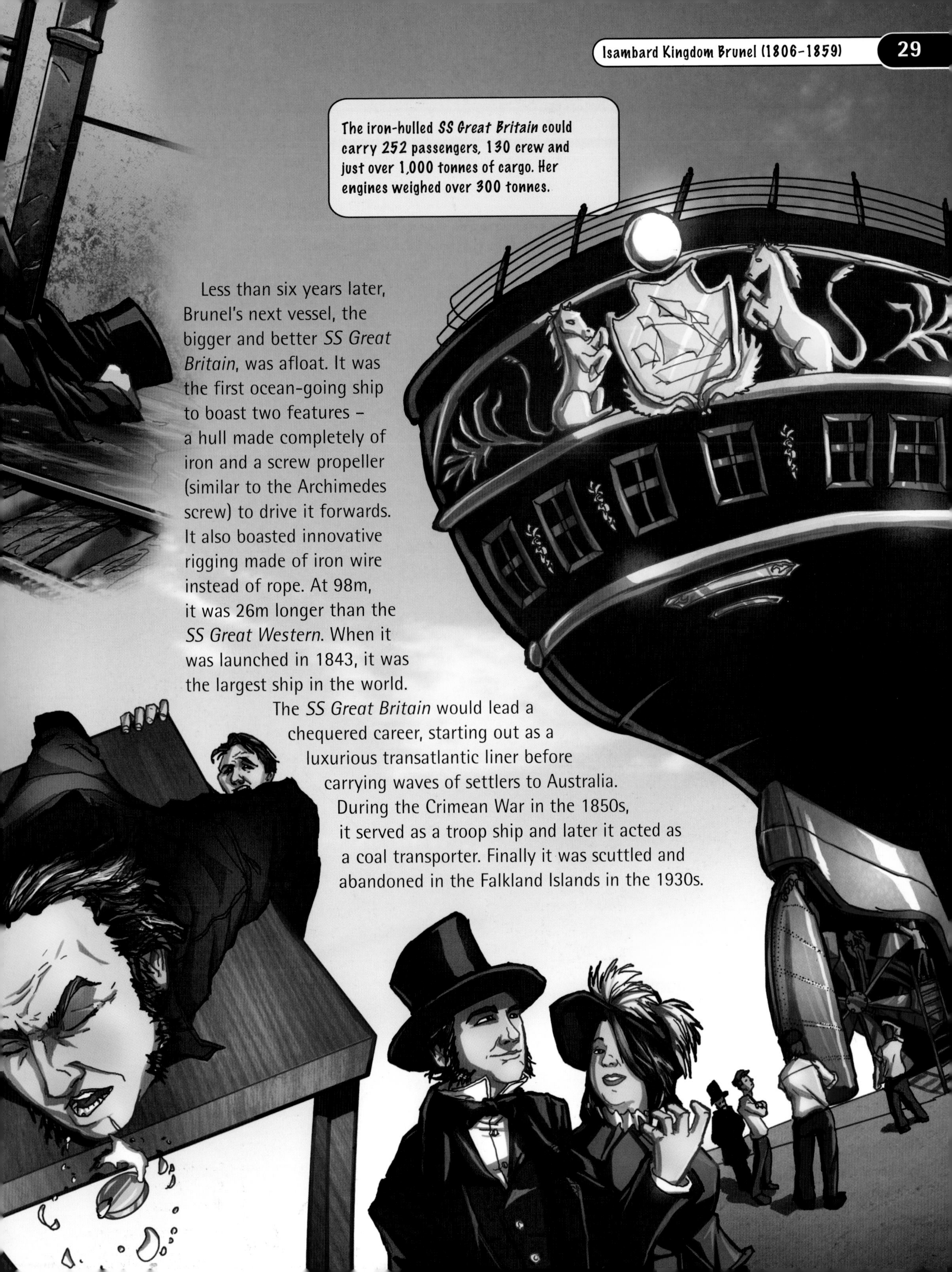

Less than six years later, Brunel's next vessel, the bigger and better *SS Great Britain*, was afloat. It was the first ocean-going ship to boast two features – a hull made completely of iron and a screw propeller (similar to the Archimedes screw) to drive it forwards. It also boasted innovative rigging made of iron wire instead of rope. At 98m, it was 26m longer than the *SS Great Western*. When it was launched in 1843, it was the largest ship in the world.

The *SS Great Britain* would lead a chequered career, starting out as a luxurious transatlantic liner before carrying waves of settlers to Australia. During the Crimean War in the 1850s, it served as a troop ship and later it acted as a coal transporter. Finally it was scuttled and abandoned in the Falkland Islands in the 1930s.

Brunel did what he could to help the war effort in Crimea, which today is part of Ukraine on the coast of the Black Sea. He designed a floating armoured barge and a series of innovative, flat-packed buildings that would hold up to 2,200 hospital beds.

Although Brunel had married Mary Horsley in 1836 and they had three children together, he was rarely at home, such was his appetite for work. But his ideas were not always successful. He set up an 'atmospheric' railway in southwest England. The trains were powered by air pressure, but the leather flaps that sealed the vacuum pipes were eaten by rats, and the railway closed within a year.

The *SS Great Eastern*, Brunel's final and biggest ship, was also not a commercial success. He began work on it in 1852 with a sketch of a vessel designed to travel to India and Australia carrying all of its own fuel. The vessel was a staggering 212m long and powered by five steam engines that drove two 17m-tall paddle wheels and a screw propeller over 7m wide. Many could not believe its sheer scale or that it could carry up to 4,000 people.

The first attempt to slide the monstrous *SS Great Eastern* sideways into the Thames was a disaster. A winch and some chains broke, killing several workers.

Financial and technical problems dogged the building of the *Great Eastern* and these were exhausting Brunel. In 1858 he was ordered to take a complete rest – although his idea of rest was touring the Swiss Alps and working on plans for the East Bengal Railway in India.

In 1859, shortly after his ship's launch, Brunel had a stroke and died ten days later. He was just 53 years old, but he left behind a great legacy. Brunel's inventing and engineering bravado had changed transport in 19th-century Britain and been influential in other parts of the world as well.

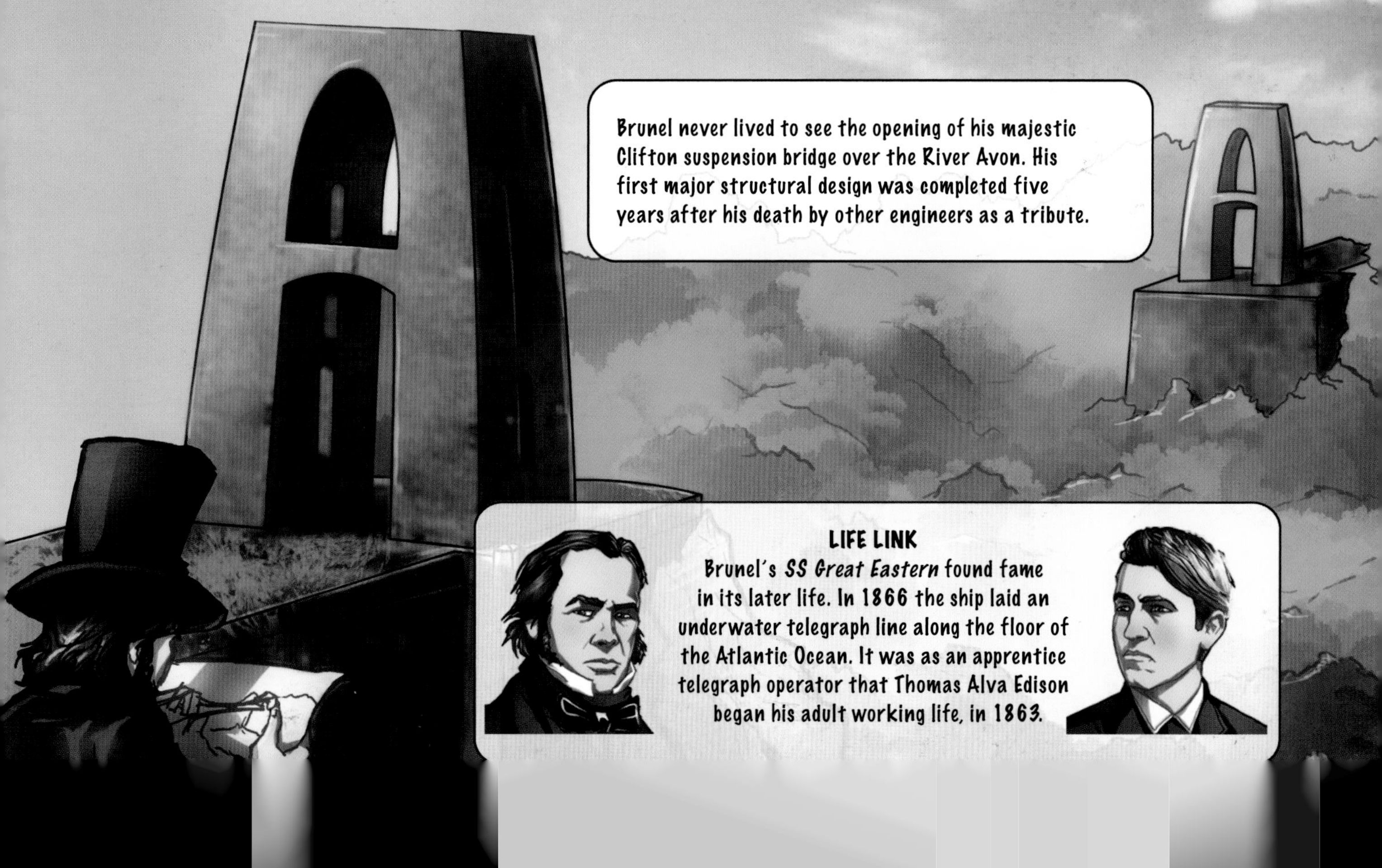

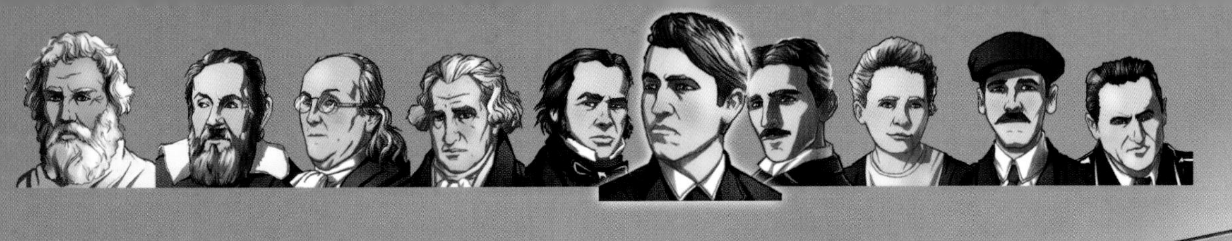

Thomas Edison

Little Jimmy had been playing with pebbles as the boxcar rolled towards him. Thomas Edison saved his life.

Michigan, USA, 1862. Two-year-old Jimmy MacKenzie wandered onto the train tracks at the Mount Clemens railway depot, totally unaware of the approaching danger. A single boxcar was rolling towards the toddler with mounting speed. There was no brakeman to stop it.

A teenage newspaper seller flung his papers down, dashed over and pulled the boy to safety just in time. Little Jimmy's father, the stationmaster, was hugely grateful and repaid the teenager's swift actions by teaching him how to use the station's telegraph system. Within weeks the boy, Thomas Alva Edison, mastered it better than most adults. Within 20 years, he would be a millionaire.

Thomas set up a 'laboratory' on the train. But when he spilt some phosphorous and started a fire, he and his equipment were thrown off the train at the very next stop.

Thomas Edison was born in Ohio and was seven when his family moved to Port Huron, Michigan. At school he was hyperactive and curious and this, together with his bad hearing, saw him labelled 'addled' (mixed up, confused) by his teacher. Thomas's mother began to home-school him and he soon developed a hunger for science, performing experiments in the basement at home.

When a railway line opened linking Port Huron to Detroit, 12-year-old Thomas began travelling on the trains, selling sweets, fruit and a newspaper that he wrote and printed himself.

As a teenager, Edison began work as a wandering 'tramp telegrapher', selling his telegraphy services to the highest bidder in Canada and northern USA. In 1867 he moved to Boston and produced his first inventions, which included an electric cockroach-killer and an automatic way of voting in elections. The voting machine was the first device he patented, meaning that no one else could copy it. It did not make him a penny, but later Edison became very rich, owning or co-owning a staggering 1,093 patents.

His first successful device was the Universal Stock Printer, which communicated the prices of stocks and shares. This and other related inventions were sold for a large sum – US$40,000. Edison sent money back to his parents and then, in 1871, he opened his own workshop in Newark, New Jersey. There he invented a telegraph printer and a 'duplex' telegraph system that could send two messages down the one wire at the same time, rather than one.

Working one night as a tramp telegrapher, Edison was meant to send out a signal every half hour. Deciding to sleep instead, he wired a clock to the telegraph machine so that the signal was sent out automatically. He was soon found out and fired.

Edison fell in love with young Mary Stilwell, who came to work at his Newark factory. They were married on Christmas Day 1871, when Mary was 16 and Thomas was 24.

Making a ticking noise, Edison's Universal Stock Printer could print one letter or number every second onto the long, thin 'ticker' tape.

Menlo Park was the first industrial research and development laboratory in the USA, if not the world. Edison worked there tirelessly, filling more than 3,000 notebooks with new ideas.

In 1876 Edison moved his workshop from Newark to Menlo Park to cut costs. His father had built him a laboratory and workshop there. Edison gathered a team of talented scientists, engineers and technicians, and he expected everyone to work as hard as himself. Their work soon paid off. Rarely a month or two went by without a new patent or invention. The first was spurred on by news of Alexander Graham Bell's telephone in 1876. Bell's device did not amplify the voice well and could be used only over short distances. Working with tiny specks of carbon, Edison created a telephone transmitter that worked much better over long distances. It would become the basis of the microphone and future telephone systems.

Edison's telephone work led him to the idea of recording sound messages. Along with his employee John Kruesi, he developed the phonograph in 1877. It recorded sound vibrations as indentations on tinfoil that could be played back. Even though the sound quality was poor, it stunned those who heard it and the phonograph made Edison famous. He became known as the 'Wizard of Menlo Park' and the 'Inventor of the Age'.

Edison's fame grew with his next major work. He didn't invent the first electric lightbulb, but he did invent practical versions of the incandescent lamp that could burn long and bright enough to be of great use in homes and buildings.

In 1878 Edison travelled to the White House in Washington to demonstrate his phonograph to President Rutherford B Hayes. The president was so amazed to hear the faint, crackling recording of a voice that he insisted his wife, Lucy, rise from her bed in the middle of the night to listen.

In 1879 Edison and British engineer Charles Batchelor created a cotton thread coated in carbon that could glow inside a bulb for a whole day. It did so as air had been sucked out of the bulb to form a vacuum.
Soon Menlo Park was lit up by strings of bulbs hanging on wires between the trees and buildings. The lights could be seen from many kilometres away and sightseers came from far and wide to see this new wonder.

With breathtaking ambition, Edison's team developed a complete system to deliver electricity to homes and offices. Edison designed a powerful type of dynamo, which converts movement into electricity, and he created fuses, circuits and other features of electrical systems that we take for granted today. In 1881 his house was the first to be fitted with a complete electric-light system, and a year later a system was set up in London to power 2,000 lamps.

Some of Edison's work was not successful. His talking doll flopped and his electric pen, which duplicated writing, sold well for only a short time. He spent much of the 1890s working on new processes to separate metals from rocky ores as well as pioneering cement and concrete construction.

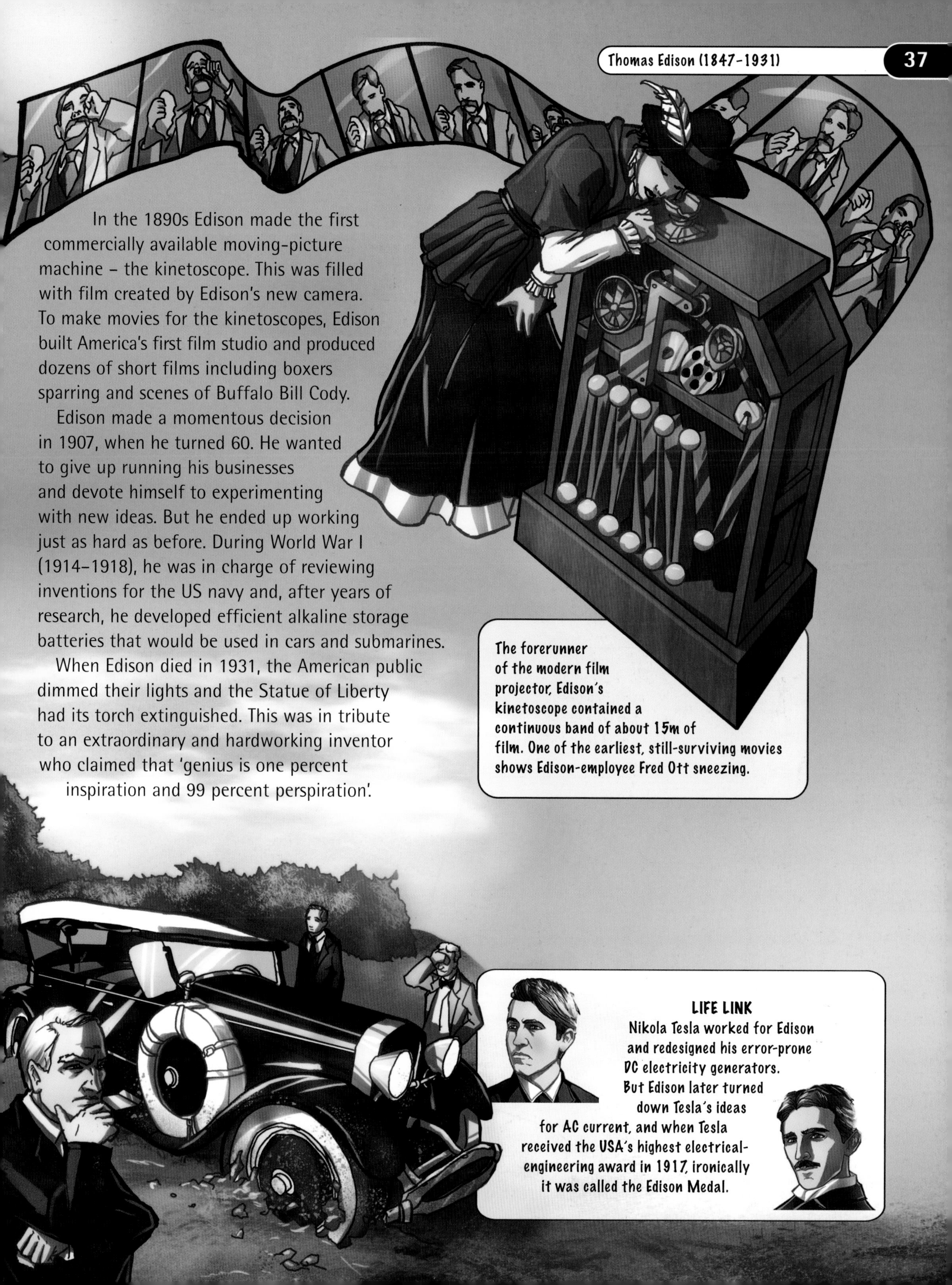

In the 1890s Edison made the first commercially available moving-picture machine – the kinetoscope. This was filled with film created by Edison's new camera. To make movies for the kinetoscopes, Edison built America's first film studio and produced dozens of short films including boxers sparring and scenes of Buffalo Bill Cody.

Edison made a momentous decision in 1907, when he turned 60. He wanted to give up running his businesses and devote himself to experimenting with new ideas. But he ended up working just as hard as before. During World War I (1914–1918), he was in charge of reviewing inventions for the US navy and, after years of research, he developed efficient alkaline storage batteries that would be used in cars and submarines.

When Edison died in 1931, the American public dimmed their lights and the Statue of Liberty had its torch extinguished. This was in tribute to an extraordinary and hardworking inventor who claimed that 'genius is one percent inspiration and 99 percent perspiration'.

The forerunner of the modern film projector, Edison's kinetoscope contained a continuous band of about 15m of film. One of the earliest, still-surviving movies shows Edison-employee Fred Ott sneezing.

LIFE LINK

Nikola Tesla worked for Edison and redesigned his error-prone DC electricity generators. But Edison later turned down Tesla's ideas for AC current, and when Tesla received the USA's highest electrical-engineering award in 1917, ironically it was called the Edison Medal.

Nikola Tesla

The man who would revolutionize the use of electricity was born just as a lightning storm engulfed his home village of Smiljan, Croatia. Nikola Tesla had a happy early childhood, but as he grew older he became awkward. After the death of his brother he began to distance himself from people, preferring to read, write poems, study nature and build simple inventions such as small water wheels.

In 1870, aged 14, Nikola left home to study at a school in Carlstadt, completing four years of schooling in three. He convinced his father not to enrol him as a priest but to let him study in Graz. There, Tesla saw a demonstration of a direct current (DC) electricity generator and his mind buzzed with ways to improve it.

Tesla went to work as a telephone engineer in Hungary and then, in 1882, he moved to France to work for Thomas Edison's company Edison Continental in Paris.

In 1883 Tesla invented the first ever motor to run on alternating current (AC). Most scientists had ignored AC electricity, seeing it as untameable, but it had a number of advantages over DC – for one, it could be transmitted over far longer distances more efficiently. But Tesla could find no buyers for his revolutionary concept. So in the spring of 1884, he sailed from Europe to the USA, hoping to meet with Thomas Edison himself. He carried with him a letter of introduction from Charles Batchelor, who was one of Edison's closest associates. The letter read: 'I know two great men and you [Edison] are one of them; the other is this young man.'

Tesla arrived in New York after a terrible journey from Paris in which he was robbed. He arrived with just four US cents in his pocket along with some poems and calculations for a fanciful flying machine.

Tesla met Edison shortly after reaching New York. Both men were intense, dedicated and hard working, but they were very different in many other ways. Tesla was studious, softly spoken and naive. Edison was shrewd, confident and a sharp businessman. He rejected Tesla's AC proposals, but took him on to improve his DC equipment. Tesla excelled himself, designing more efficient equipment for Edison and fixing generators on the *Oregon* cruise ship, but the two fell out when Edison refused Tesla's demands for payment. Tesla resigned.

Tesla claimed that Edison offered to pay him $50,000 for redesigning some motors and generators. When Tesla asked for the money in 1885, Edison backed down on his promise, saying, 'Tesla, you don't understand our American humour.'

Tesla earnt money by digging ditches and working as a labourer, but he never stopped developing his ideas. He invented a complete system, including all its parts – dynamos, transformers and motors – to generate and transmit AC electricity over long distances. This astonishing achievement changed the future of electricity. George Westinghouse's electrical company paid Tesla handsomely for the idea. He received one million US dollars upfront and generous future royalties on every unit of electricity sold – a sum likely to total tens of millions.

Westinghouse and Tesla found themselves pitted in a 'war of the currents' with Edison and his DC electricity system, which could transport electricity only 1 or 2km. Grand demonstrations of AC electricity by Tesla and Westinghouse gradually showed its value, and AC is now the way electricity is distributed to most of the world's population.

Tesla invented the brushless induction motor, which ran on AC electricity, in 1887. His AC motors were faster, simpler and more reliable than DC motors of the time.

Tesla and Westinghouse's AC electricity system lit an incredible 96,620 lamps at the 1893 Chicago World's Fair. The 'City of Light' was seen by over 25 million people and proved a major turning point in the war of the currents.

Tesla ploughed almost all of his new-found wealth into a laboratory in New York. This became home to many astonishing accomplishments. Tesla invented new, brighter forms of lights and experimented with sending electricity through the air to light lamps without wires. He also invented the Tesla coil, an electrical device that would be used in many early radio and TV sets. Tesla found that his coils could be used to send and receive radio waves. He made this discovery several years before Guglielmo Marconi, often considered to be the inventor of radio, began his experiments.

The year 1895 was one of triumph and trouble. A major hydro-electric power station using Tesla's generators began operating at Niagara Falls, providing power the following year to the large city of Buffalo, 35km away. Tesla had also been preparing to stun the world with his radio demonstrations and other work. Then disaster struck – a fire raged through the building housing his laboratory. It destroyed Tesla's work.

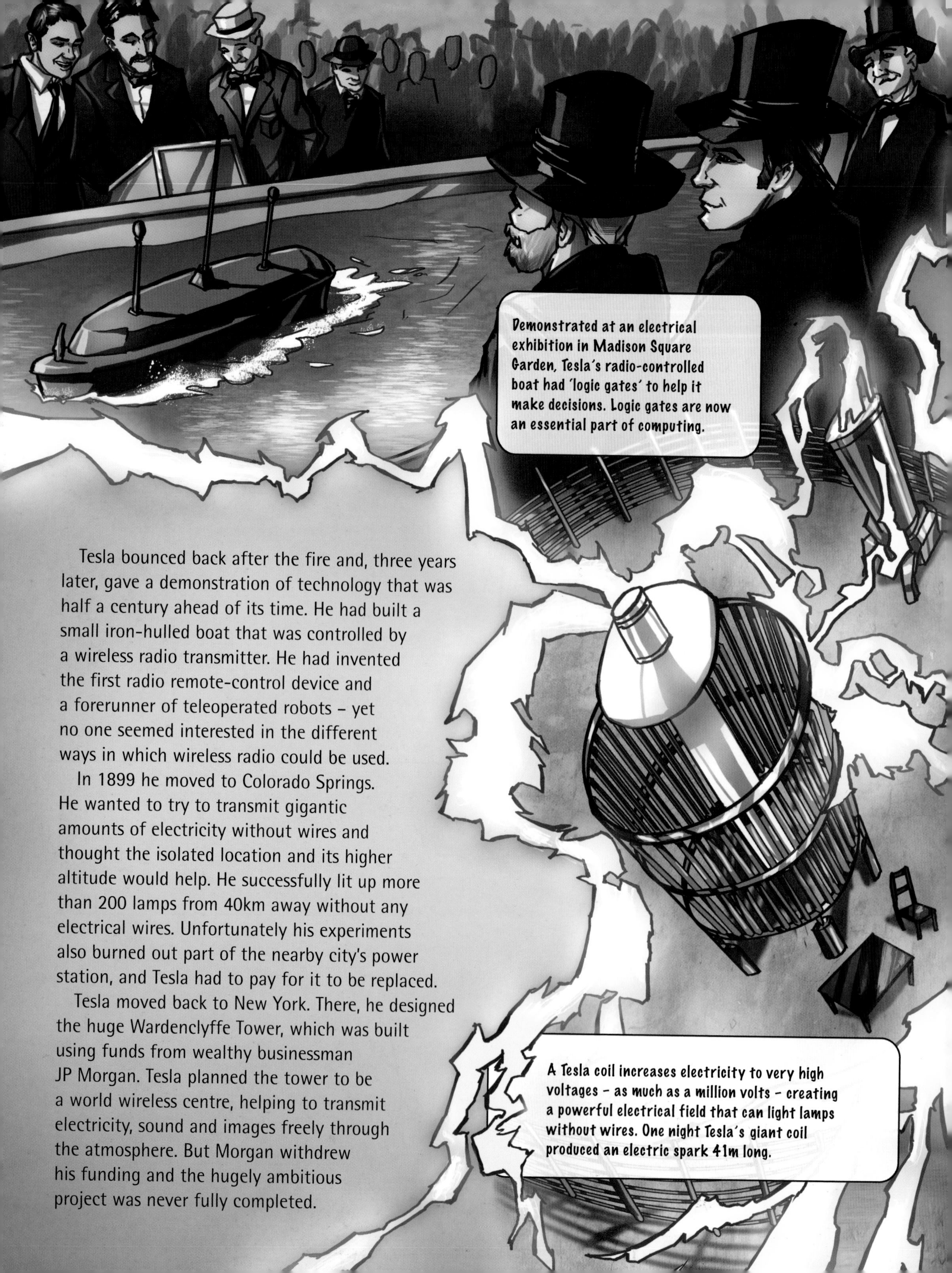

Tesla bounced back after the fire and, three years later, gave a demonstration of technology that was half a century ahead of its time. He had built a small iron-hulled boat that was controlled by a wireless radio transmitter. He had invented the first radio remote-control device and a forerunner of teleoperated robots – yet no one seemed interested in the different ways in which wireless radio could be used.

In 1899 he moved to Colorado Springs. He wanted to try to transmit gigantic amounts of electricity without wires and thought the isolated location and its higher altitude would help. He successfully lit up more than 200 lamps from 40km away without any electrical wires. Unfortunately his experiments also burned out part of the nearby city's power station, and Tesla had to pay for it to be replaced.

Tesla moved back to New York. There, he designed the huge Wardenclyffe Tower, which was built using funds from wealthy businessman JP Morgan. Tesla planned the tower to be a world wireless centre, helping to transmit electricity, sound and images freely through the atmosphere. But Morgan withdrew his funding and the hugely ambitious project was never fully completed.

The war of the currents had raged on. AC electricity won but at great cost to the companies promoting it. Far more a scientist than businessman, Tesla accepted the sum of US$216,000 from Westinghouse instead of the millions owed in royalties. He began to struggle with money and, in 1916, he declared himself bankrupt. Most of his extraordinary ideas stayed in his notebooks due to a lack of funds. These included using radio waves to detect submarines (similar to radar) and a powerful particle ray, a little like a laser, that could be used as a weapon. Many scoffed at his ideas, but the FBI built up a 156-page file on him and may have confiscated his papers after his death.

Tesla became more eccentric as he grew older. He was obsessed with his own personal hygiene, loathed touching certain items and would stay only in hotel rooms whose numbers could be divided by three. In 1943 Tesla died alone and penniless in his suite in the Hotel New Yorker. He left behind an incredible legacy. Shortly after his death, the US Supreme Court effectively declared that Tesla, not Marconi, was the first inventor of radio. More than anything, Tesla was an endlessly inventive and curious genius who was often years ahead of his time.

For the last ten years of his life, Tesla lived in Room 3327, a two-room suite on the 33rd floor of the Hotel New Yorker. There, his most frequent visitors were pigeons, which he fed and befriended.

Tesla's huge Wardenclyffe Tower was never operational and newspapers laughed at his expensive mistake. It was torn down in 1917 as the US government feared it might be used by enemy spies.

LIFE LINK

During his work into high-frequency electricity, Tesla came across the phenomenon of X-rays, which he called 'shadowgraphs', and he made images of his own body. Marie Curie's work helped X-rays become a vital tool in medicine, and she equipped and trained hundreds of X-ray units during World War I.

Marie Curie

Paris, 1898. Inside a damp, bare, unheated shed, Marie Curie had been working long, hard hours. She was investigating a mineral called pitchblende, trying to find the source of its powerful particles and rays of energy that we know today as radiation. Finally, she discovered a new chemical element. Curie had already invented the word 'radioactivity', and now she named the new element radium.

Born in Warsaw, Poland, Marie Sklodowska boasted an excellent memory as a child and learned to read aged four. Poland was under Russian control and so Marie's lessons were in Russian, but she still excelled. Like her elder sister, Bronia, and her brother, Joseph, she won the gold medal for best pupil at her secondary school. At 16, she began teaching to raise money to send Bronia to university in Paris. It took eight years before, in 1891, Marie could finally afford to study in Paris too.

At the famous Sorbonne, the University of Paris, Marie threw herself into her studies in mathematics and physics. She rarely had the money or the interest to take part in Paris's stylish lifestyle, concentrating instead on her physics and maths degrees. In 1894, while researching magnetism in different steels, she met Pierre Curie, a 35-year-old scientist who had won acclaim for his work on magnetism and crystals.

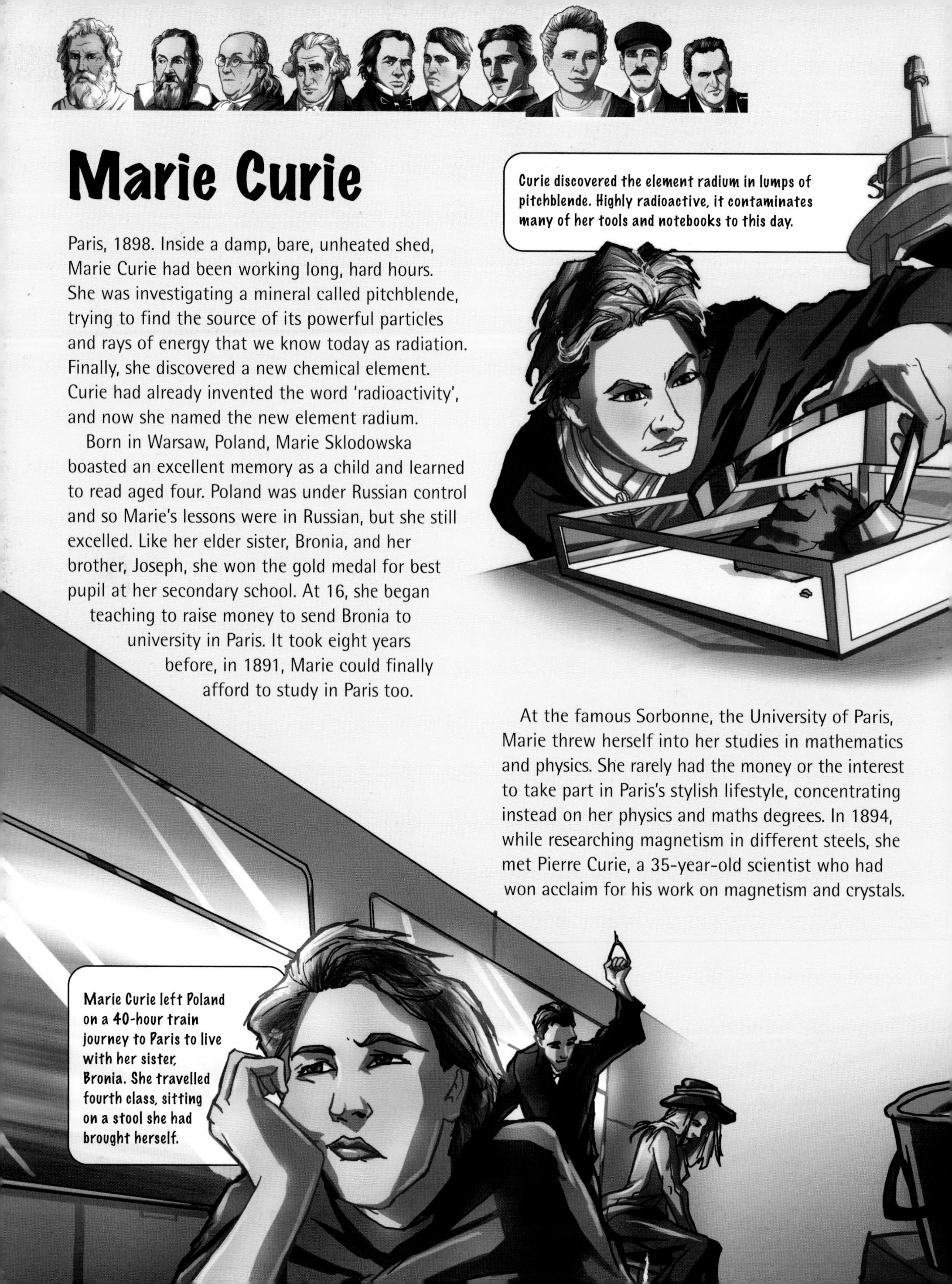

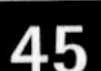

In 1896 Henri Becquerel discovered mysterious rays coming from uranium. News reached the Curies fast and Marie switched research to investigate whether this 'radiation' existed in other elements.

Marie and Pierre discovered that their main source of uranium, an ore called pitchblende, possessed much higher levels of radioactivity than pure uranium. Something else in the pitchblende must be highly radioactive! Pitchblende is made up of over 25 chemical elements and refining it to isolate each element proved long, hard work. Finally in 1898, they discovered not one but two new elements: polonium and radium. Polonium was impossible to isolate totally, but radium looked more promising. They bought many tonnes of industrial-waste pitchblende, and Marie spent four years isolating just one tenth of a gram of radium. Yet even this tiny amount was enough. Pierre and Marie discovered that radium was about a million times more radioactive than uranium.

Pierre and Marie married in 1895. Their unusual honeymoon was a cycle ride around the French countryside.

The clamour for radium grew and the Curies were becoming famous, a fate sealed when they were awarded, with Henri Becquerel, the 1903 Nobel Prize for Physics. Marie was the first woman to win the prestigious award. She and Pierre were too sick to travel to Sweden to receive the prize. They were probably suffering from the effects of radiation exposure, a danger not fully understood at the time.

Refining pitchblende was arduous work. The ore had to be ground down, sieved, boiled and distilled before going through a process called electrolysis. Curie's laboratory was a large, ramshackle shed with a glass roof that barely kept out the rain.

Tragedy struck in 1906, when Pierre was killed. His death came as a shocking blow to Marie, who was left with two young daughters, Irene and Eve, to look after, but she vowed to continue working.

Curie investigated radium's potential in medicine. Radium gave off radon gas, which could be collected, sealed in tubes and then put to use to destroy cancer tumours.

While radium could be used combined with other chemicals in substances called radium salts, Marie managed to isolate pure radium so that she could study all of its properties. This and her other work saw her become the first person to win a second Nobel Prize, this time for Chemistry in 1911.

In 1909 the Sorbonne and the Pasteur Institute decided to finance a Radium Institute. Curie was excited to be getting a purpose-built workplace. But by the time the building was ready in 1914, World War I had begun. Most of Curie's staff had volunteered or been enlisted for the French army.

Curie knew that X-rays could be vital to treat wounded soldiers by finding bullets and metal shrapnel hidden in their bodies. She campaigned for funds, scavenged equipment and began fitting out hospitals with X-ray facilities.

Curie set up many mobile X-ray vehicles, known as 'Petites Curies'. She also organized about 200 facilities in field hospitals, training hundreds of staff to operate them. During the war, these facilities X-rayed over a million wounded, saving many lives.

The Radium Institute in Paris was now open, but it lacked resources. In 1920 an interview with journalist Maria Meloney would become crucial to its future. Meloney was shocked to learn that the discoverer of radium had just one gram of the element for her entire institute to work with, while the USA had 50 times that amount. Back in America, Meloney managed to raise US$100,000 to buy radium for the institute. Marie made her first trip across the Atlantic in 1921 to receive the radium as well as money and useful equipment.

During the 1920s, Curie continued to work at the institute, researching the peaceful, medical uses of radioactivity. She was joined by Irene and, in 1925, by an assistant, Frédéric Joliot, who married Irene the following year.

The decades of exposure to harmful radiation with little or no protection began to take their toll. Curie became increasingly ill, but she refused to retire. In 1934, just as Irene and Frédéric were making major breakthroughs in turning a non-radioactive substance into a radioactive one, Curie died. The year after her death, Irene and Frédéric Joliot-Curie were awarded the Nobel Prize. Marie and Irene became the first mother and daughter to both be Nobel winners.

In a society where male scientists dominated, Marie Curie stood out for her dedication, intelligence and strong will. She helped build a path for other female scientists to follow.

Glenn Curtiss

January 1907. Glenn Hammond Curtiss sat low over the frame of his monstrous motorcycle and raced along Ormond Beach in Florida, USA. The engine had screamed into life after a push start from friends Tank Walters and Thomas Baldwin. They feared for him. No one had travelled this fast before – and his vehicle had no brakes. Most motorcars of the time could reach 60 or 70km/h at best. Curtiss's speed was astounding – 219km/h. He was the fastest person on Earth.

Curtiss grew up in New York State. His father died when he was five and at 14, he worked at the Eastman Kodak Company, assembling cameras. He built up a reputation as an able mechanic who could repair almost any device, and in 1900, he opened his first bicycle shop.

Curtiss's 1907 record-breaking motorbike was much longer than a regular machine as it needed room for his own huge V8 engine.

The following year, Curtiss attached an engine to a bike frame for the first time. The engine was sent by mail order, and Curtiss was disappointed with its quality, so he began tinkering with it to improve it. His first carburettor (the part of an engine that blends together air and fuel) was made of an old tomato can stuffed with wire wool.

The teenaged Curtiss bought an old bicycle, fixed it up and got a job as a courier for Western Union, delivering packages. In his spare time, Curtiss picked up extra dollars as a racing cyclist.

Curtiss started racing his homemade motorbikes and set a speed record of 103km/h over a mile (1.6km). Orders began to pour in for his engines and motorbikes, which were made under the Hercules brand. Within a few years, Curtiss was regarded as the best maker of lightweight engines in the USA. One of his engines was sold to a former circus-trapeze artist, Thomas Baldwin, who fitted it to his *California Arrow* airship in 1904. It was in this machine that Curtiss made his first flight.

Telephone inventor Alexander Graham Bell was very keen to work with Curtiss. The two men formed the Aerial Experiment Association (AEA), which made six aircraft. In 1908 their third craft, *June Bug*, designed by Curtiss, made the first public flight over 1km in the USA, winning a large prize. The following year, Curtiss designed a new aircraft with long ailerons (flaps for helping aircraft turn) and a water-cooled engine. In this machine, Curtiss beat famed French aviator Louis Bleriot over a 20km course to win the Gordon Bennett Cup.

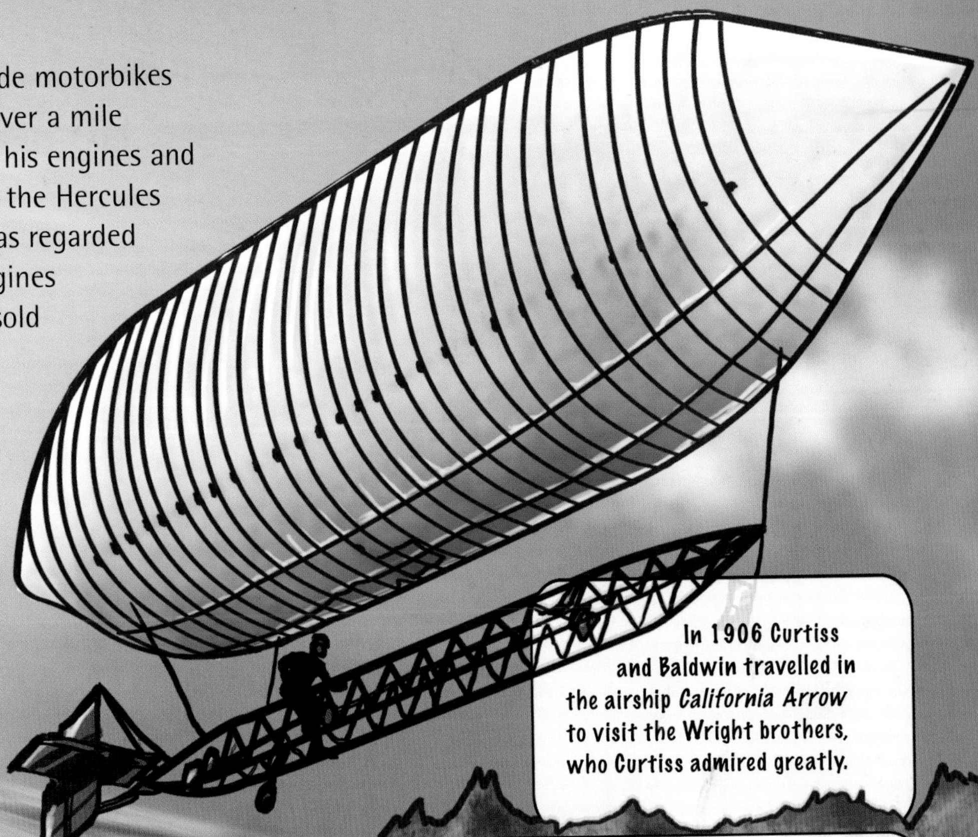

In 1906 Curtiss and Baldwin travelled in the airship *California Arrow* to visit the Wright brothers, who Curtiss admired greatly.

The meeting with Orville and Wilbur Wright was friendly, but the brothers showed little interest in working with Curtiss.

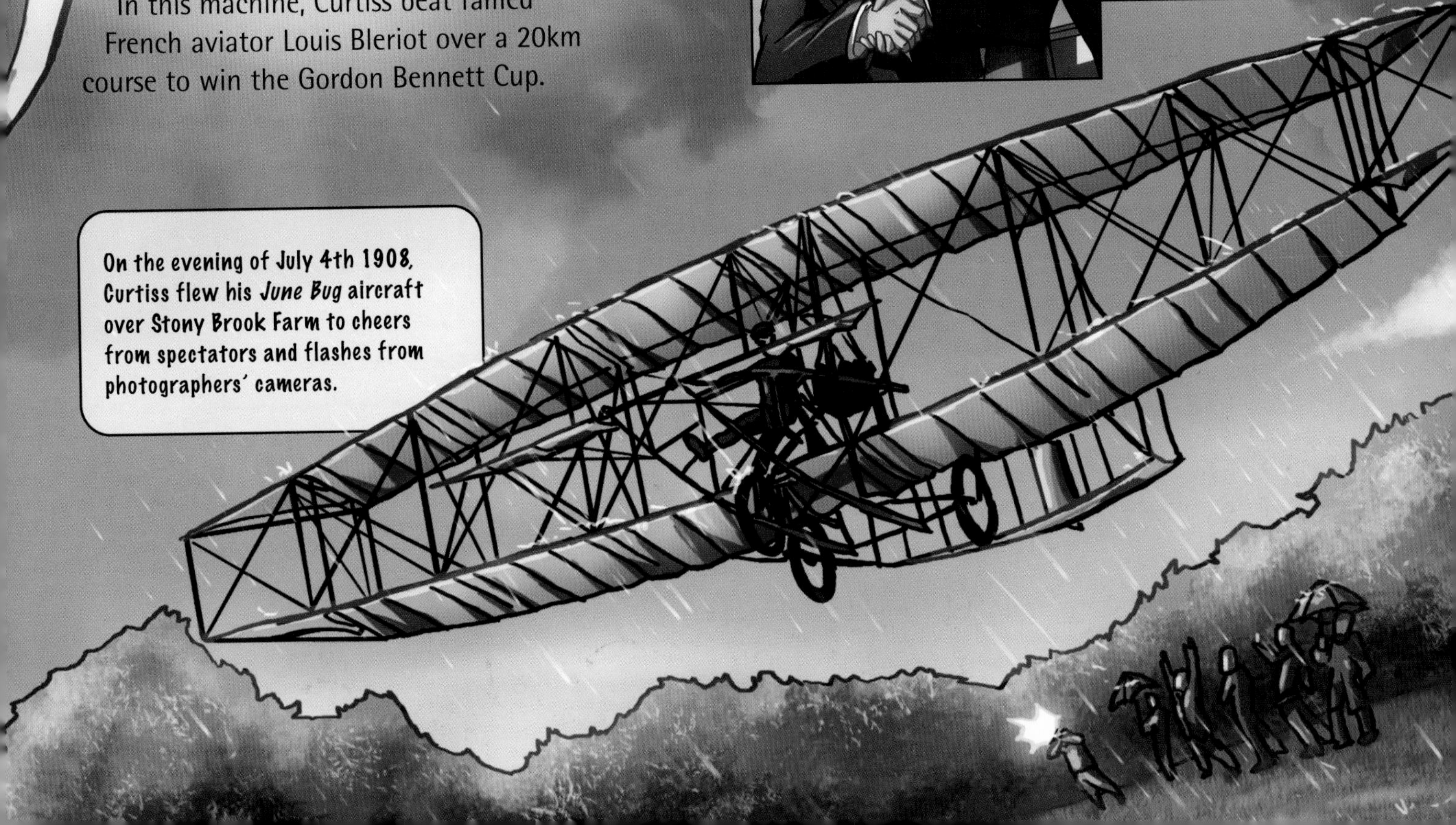

On the evening of July 4th 1908, Curtiss flew his *June Bug* aircraft over Stony Brook Farm to cheers from spectators and flashes from photographers' cameras.

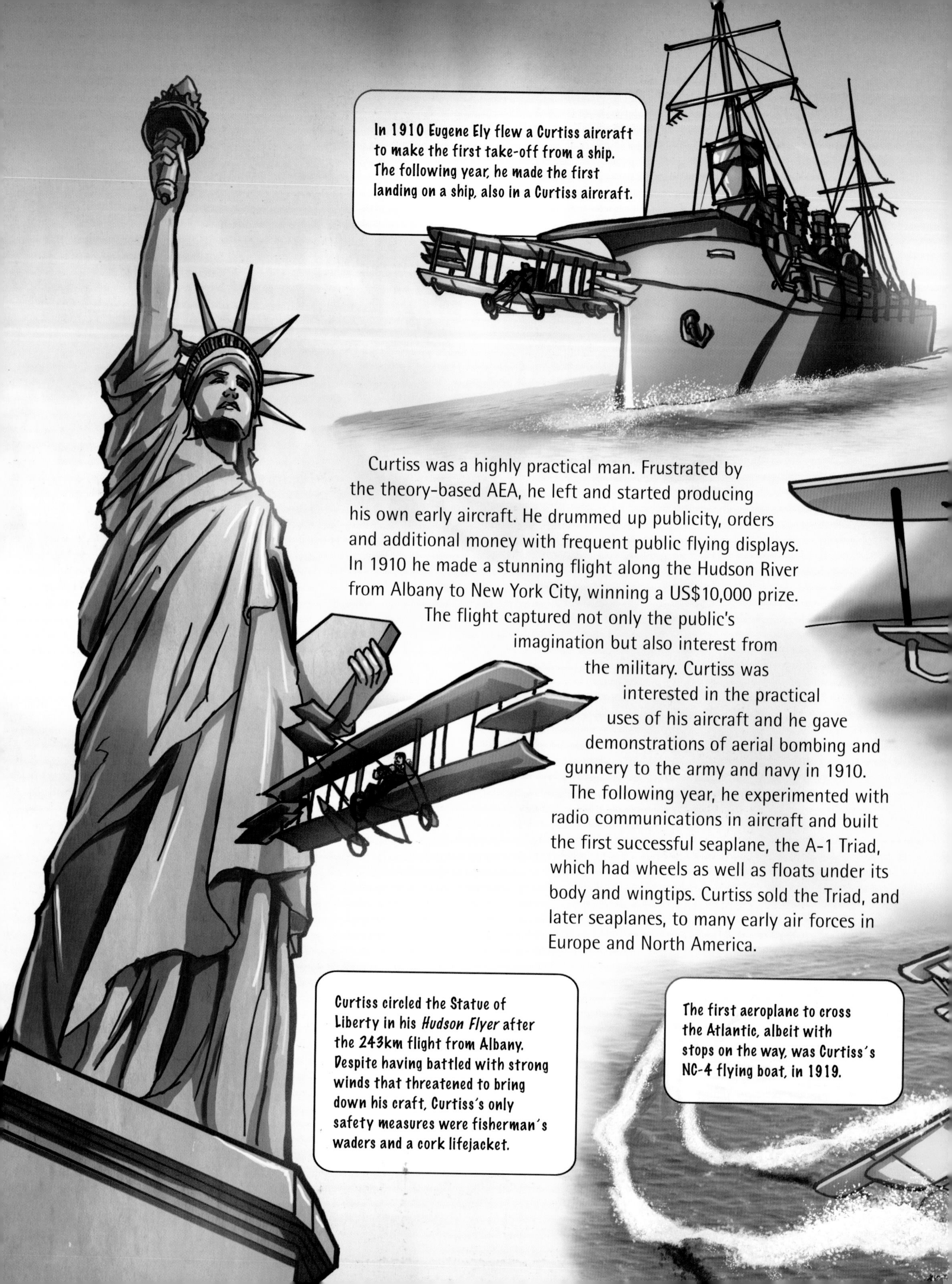

Curtiss was a highly practical man. Frustrated by the theory-based AEA, he left and started producing his own early aircraft. He drummed up publicity, orders and additional money with frequent public flying displays. In 1910 he made a stunning flight along the Hudson River from Albany to New York City, winning a US$10,000 prize. The flight captured not only the public's imagination but also interest from the military. Curtiss was interested in the practical uses of his aircraft and he gave demonstrations of aerial bombing and gunnery to the army and navy in 1910. The following year, he experimented with radio communications in aircraft and built the first successful seaplane, the A-1 Triad, which had wheels as well as floats under its body and wingtips. Curtiss sold the Triad, and later seaplanes, to many early air forces in Europe and North America.

Curtiss's company expanded to become the biggest aircraft maker of World War I with some 18,000 staff producing over 10,000 aircraft. Most popular of all the models was the Curtiss JN, or 'Jenny'. It was used to train thousands of military and civilian pilots, including Charles Lindbergh, who in 1927 made the first solo non-stop flight across the Atlantic from New York to Paris, France.

Curtiss left the aviation business in 1921, although he kept giving advice to aircraft makers. He moved to Florida where he made a second fortune in property. After an operation in 1929, Curtiss died aged 52. The demon racer had pioneered naval aviation and helped to invent many of the features found on modern aircraft, from dual pilot control for training aircraft to retractable landing-gear wheels.

LIFE LINK

Curtiss was the father of naval aviation and in 1911, he produced the A-1 Triad seaplanes, which were bought by the Russian military. As a teenager, Sergei Korolev's first encounter with aircraft was with WWI Russian seaplanes in which he took his first flight. He then moved into a career in aviation and space travel.

Sergei Korolev

April 12th 1961. 'Poyekhali!' ('Off we go!') spoke Yuri Gagarin over the radio as the giant Vostok launch rocket blasted off from Baikonur Cosmodrome in Kazakhstan. Within minutes, Gagarin was the first person in space. In his 108-minute-long mission, Gagarin ate, drank water and experienced weightlessness with his pencil floating just out of his reach. He remarked at the beauty of the breathtaking view from outer space.

Gagarin fired his ejection seat 7,000m above Earth and parachuted to safety. At a lavish parade and ceremony in Moscow, Russia, Gagarin was given a hero's welcome, yet the man largely responsible for the mission, the man who had designed the rocket and co-designed the Vostok space capsule, was kept on the sidelines. For his entire life, he was known to the outside world only as the mysterious 'Chief Designer'. He was Sergei Pavlovich Korolev.

Gagarin's *Vostok 1* space capsule was blasted into space by a huge rocket with five engines.

As an aircraft-obsessed teenager, Korolev thought nothing of swimming across the freezing waters of a harbour to get a closer look at a Russian squadron of seaplanes.

A Russian farmer and her daughter watched in fear as Gagarin parachuted to Earth. He told them, 'Don't be afraid. I am a Soviet like you... I must find a telephone to call Moscow!'

Korolev was just six years old when he saw the aerial antics of early Russian pilot Sergei Utochkin in his hometown of Nizhyn. It made a lasting impression on him. By the time he moved to Odessa in 1917, Korolev was hooked on aviation. He joined a local aeronautics club and started to design and build his own gliders. In 1924 he went to live with his uncle to study at the Kiev Polytechnic Institute and later he attended a technical school in Moscow, where one of his course advisors was the famous aircraft designer Sergei Tupolev. Far from wealthy, Korolev did odd jobs such as selling newspapers and working as a carpenter.

Korolev turned to rockets and space in the late 1920s. He read the works of space pioneer Konstantin Tsiolkovsky and got the chance to meet him in 1929. While working at a government aircraft-design bureau, he joined a group of young rocket engineers known as GIRD. In 1933 he was their chief designer when they launched Russia's first liquid-fuelled rocket, the 09. The military had begun to notice Korolev and soon the young designer was made deputy chief of the Jet Propulsion Research Institute (RNII).

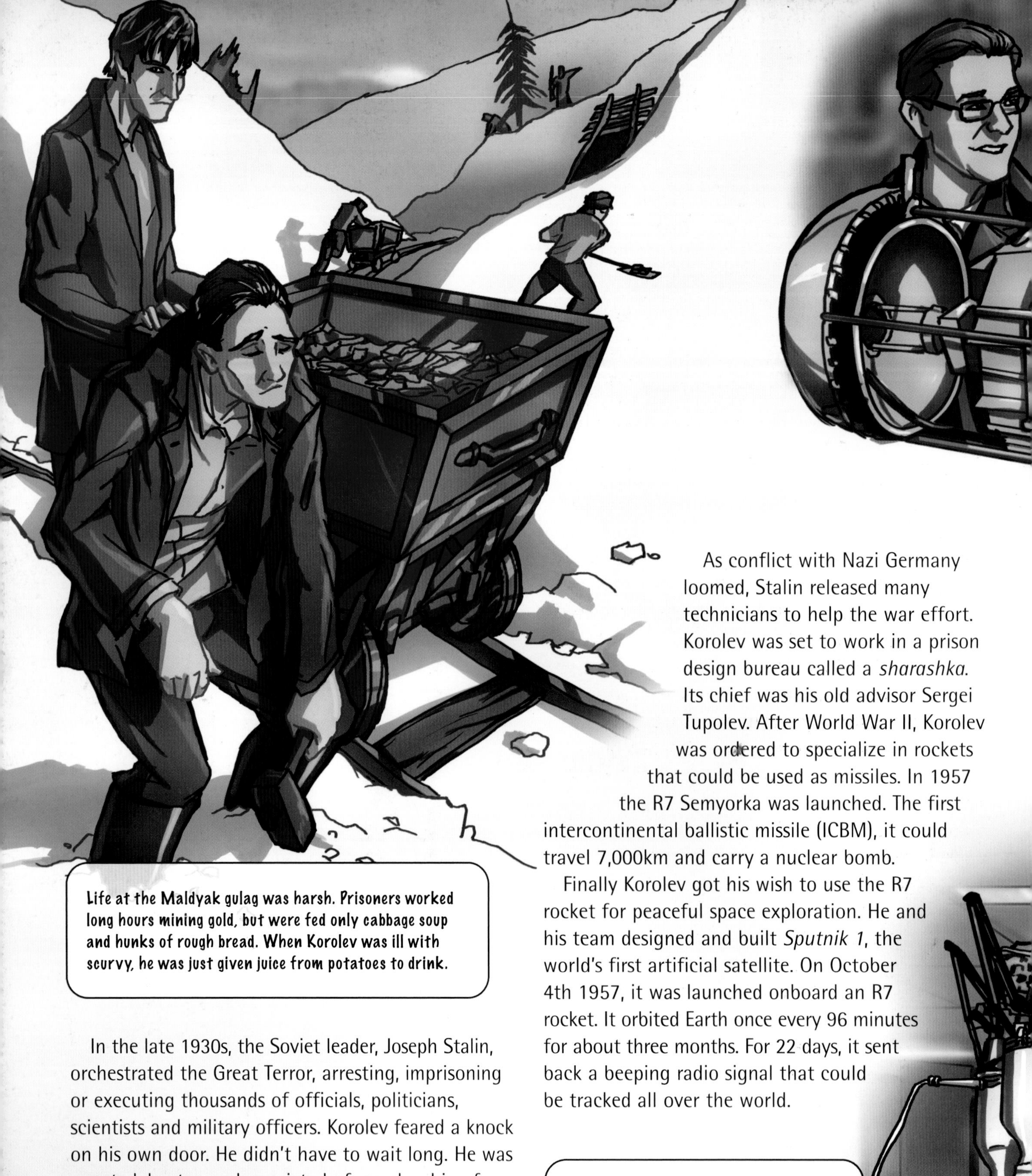

Life at the Maldyak gulag was harsh. Prisoners worked long hours mining gold, but were fed only cabbage soup and hunks of rough bread. When Korolev was ill with scurvy, he was just given juice from potatoes to drink.

In the late 1930s, the Soviet leader, Joseph Stalin, orchestrated the Great Terror, arresting, imprisoning or executing thousands of officials, politicians, scientists and military officers. Korolev feared a knock on his own door. He didn't have to wait long. He was arrested, beaten and convicted of membership of anti-Soviet groups and of sabotage. The charges were untrue, but he was sent to a *gulag*, a forced-labour camp, deep in isolated eastern Siberia. Conditions were brutal and Korolev suffered terribly, losing most of his teeth, being beaten and developing scurvy.

As conflict with Nazi Germany loomed, Stalin released many technicians to help the war effort. Korolev was set to work in a prison design bureau called a *sharashka*. Its chief was his old advisor Sergei Tupolev. After World War II, Korolev was ordered to specialize in rockets that could be used as missiles. In 1957 the R7 Semyorka was launched. The first intercontinental ballistic missile (ICBM), it could travel 7,000km and carry a nuclear bomb.

Finally Korolev got his wish to use the R7 rocket for peaceful space exploration. He and his team designed and built *Sputnik 1*, the world's first artificial satellite. On October 4th 1957, it was launched onboard an R7 rocket. It orbited Earth once every 96 minutes for about three months. For 22 days, it sent back a beeping radio signal that could be tracked all over the world.

Korolev's R7 rocket was made up of one giant rocket motor and four strap-on boosters. Together they generated vast amounts of thrust.

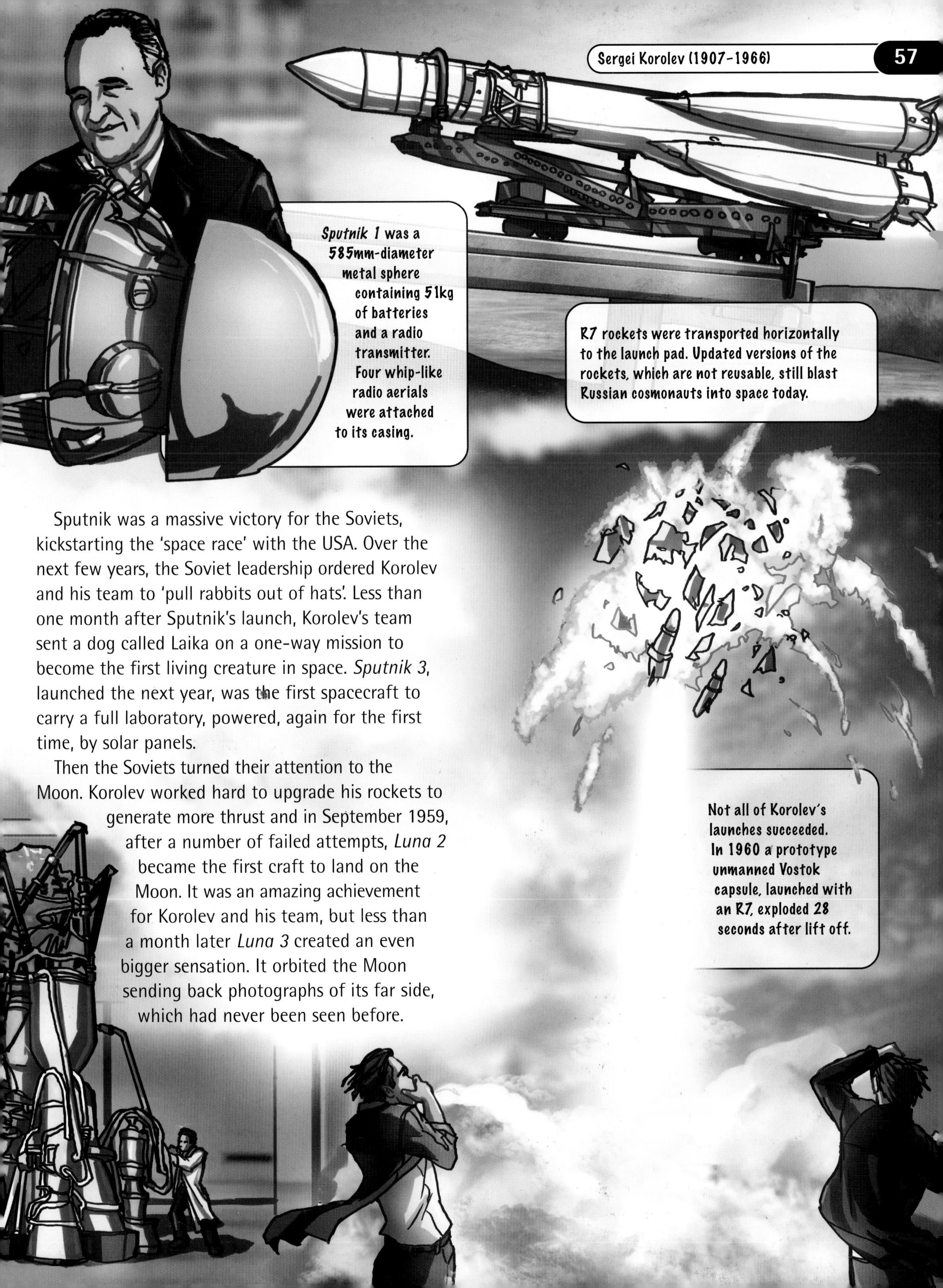

Sputnik was a massive victory for the Soviets, kickstarting the 'space race' with the USA. Over the next few years, the Soviet leadership ordered Korolev and his team to 'pull rabbits out of hats'. Less than one month after Sputnik's launch, Korolev's team sent a dog called Laika on a one-way mission to become the first living creature in space. *Sputnik 3*, launched the next year, was the first spacecraft to carry a full laboratory, powered, again for the first time, by solar panels.

Then the Soviets turned their attention to the Moon. Korolev worked hard to upgrade his rockets to generate more thrust and in September 1959, after a number of failed attempts, *Luna 2* became the first craft to land on the Moon. It was an amazing achievement for Korolev and his team, but less than a month later *Luna 3* created an even bigger sensation. It orbited the Moon sending back photographs of its far side, which had never been seen before.

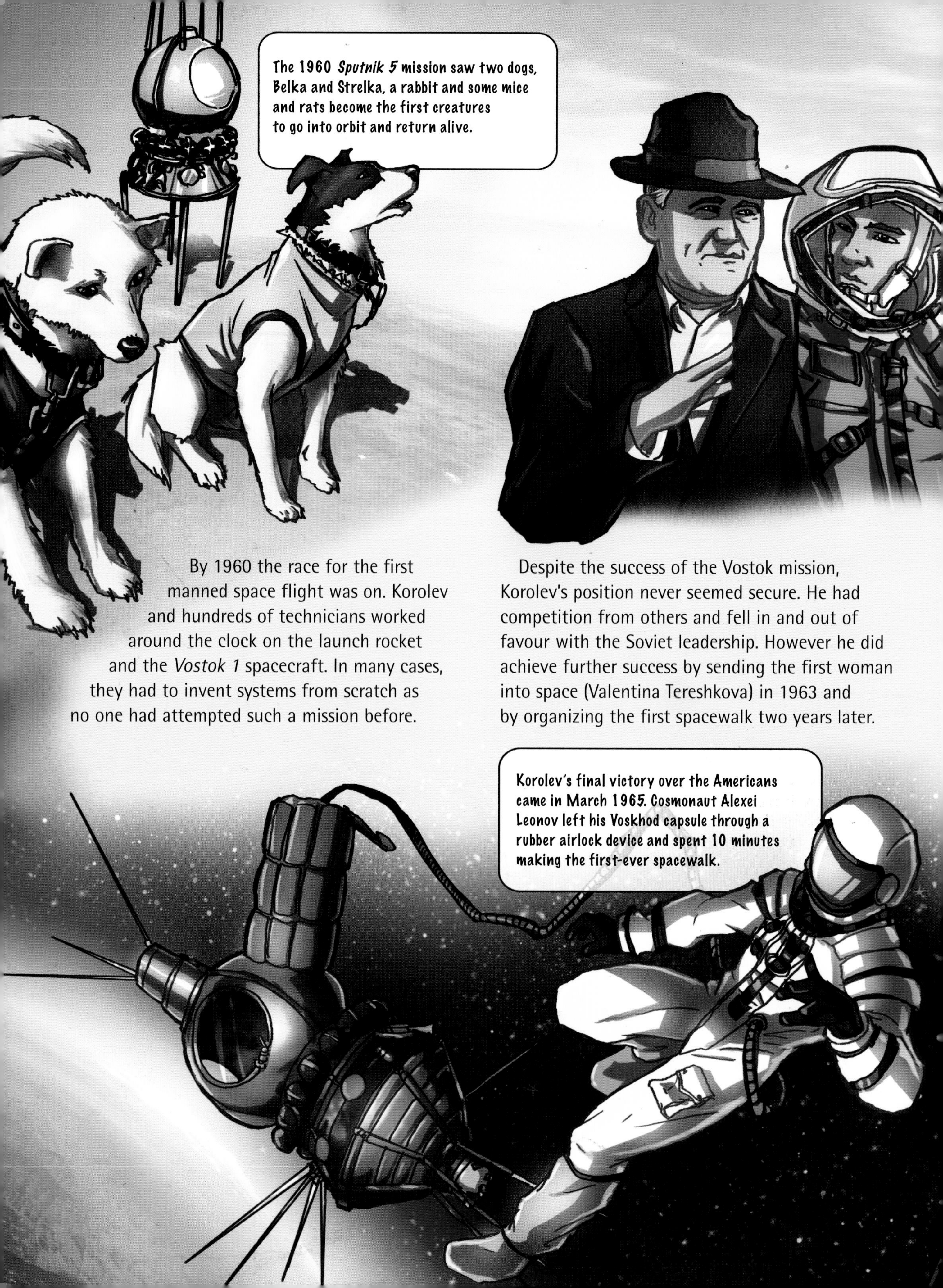

By 1960 the race for the first manned space flight was on. Korolev and hundreds of technicians worked around the clock on the launch rocket and the *Vostok 1* spacecraft. In many cases, they had to invent systems from scratch as no one had attempted such a mission before.

Despite the success of the Vostok mission, Korolev's position never seemed secure. He had competition from others and fell in and out of favour with the Soviet leadership. However he did achieve further success by sending the first woman into space (Valentina Tereshkova) in 1963 and by organizing the first spacewalk two years later.

Korolev had a major influence on the training and selection of cosmonauts. He formed a close friendship with Yuri Gagarin, who was chosen for the first manned space flight.

Improved versions of the R7 rocket continued to serve the Soviets' groundbreaking space missions, but again Korolev looked ahead, this time to manned Moon landings. He began developing a massive rocket, the N-1, and the advanced Soyuz spacecraft. Too little funding and too much political infighting hampered these projects and the Soviets would never get to the Moon. By the mid-1960s, Korolev was an ill man. He was overworked and stressed, but it was still a shock to his friends when he died shortly after a routine operation in 1966.

The Chief Designer's identity was revealed to the world only after his death and even today, he is still not that well known. Yet Sergei Korolev's vision, persistence and brilliance gave the world many breakthroughs in space technology, including the first satellite, first man and first woman in space – achievements that still echo around the world today.

Korolev designed and developed early versions of the Soyuz spacecraft, but died before he could see them become successful. The craft are still in use today.

Other famous inventors

Al Jazari (c.1136–1210)
A craftsman and engineer from northern Mesopotamia (now Iraq and Syria), Al Jazari wrote *The Book of Knowledge of Ingenious Mechanical Devices.* Published in 1206CE, it described in detail 50 extraordinarily advanced machines, such as the first-known combination lock and crankshaft as well as versions of suction pumps, mechanical clocks powered by water, and moving model figures. Many of its machines and techniques were centuries ahead of European science and engineering.

Joseph Montgolfier (1740–1810) and Jacques Montgolfier (1745–1799)
The Montgolfier brothers were sons of a wealthy French papermaker. Both were fascinated by the possibilities of flight and experimented with filling containers made of paper with hot air from a fire. In September 1783, a balloon built by the brothers took to the air carrying a duck, a sheep and a rooster to a height of over 450m. Two months later, a larger balloon, some 23m tall and 14m wide, flew two men 900m above Paris for a distance of nine kilometres – the first successful flight by humans.

Alessandro Volta (1745–1827)
Born in Como, Italy, Volta became professor of experimental physics at the University of Pavia in 1779. He investigated chemicals and electricity and in 1800, invented the Voltaic Pile – the first battery. It was made up of discs of copper, zinc and cardboard soaked in saltwater. Chemical reactions between the different materials produced a constant stream of electricity. The electrical unit, the volt, is named after him.

Clarence Birdseye (1886–1956)
After quitting college to become a naturalist, Birdseye worked in the icy wastes of Labrador, Canada. The American discovered that fish caught here froze almost immediately and when thawed out, still tasted fresh. With just a handful of dollars, Birdseye began experimenting with methods of packing fresh foods into waxed cardboard boxes, which were then frozen under pressure. The Birdseye brand began selling his wares in 1930. Birdseye also invented an infrared heater and a way of making paper pulp out of sugarcane waste.

Douglas Engelbart (born 1925)
A navy technician during World War II, Engelbart joined the Stanford Research Institute in the 1950s. There, he was part of a team that developed many pioneering technologies that we still use whenever we switch on a personal computer. Engelbart developed the NLS, or oN-Line System – a way of computers working with each other. It included the first hyperlinks (underlined text that, when clicked on, takes the user to another place) and the first computer mouse. In 1968 Engelbart showcased his inventions in a demonstration that also saw pioneering uses of video-conferencing and email. Now in his eighties, he still works at his own Bootstrap Institute.

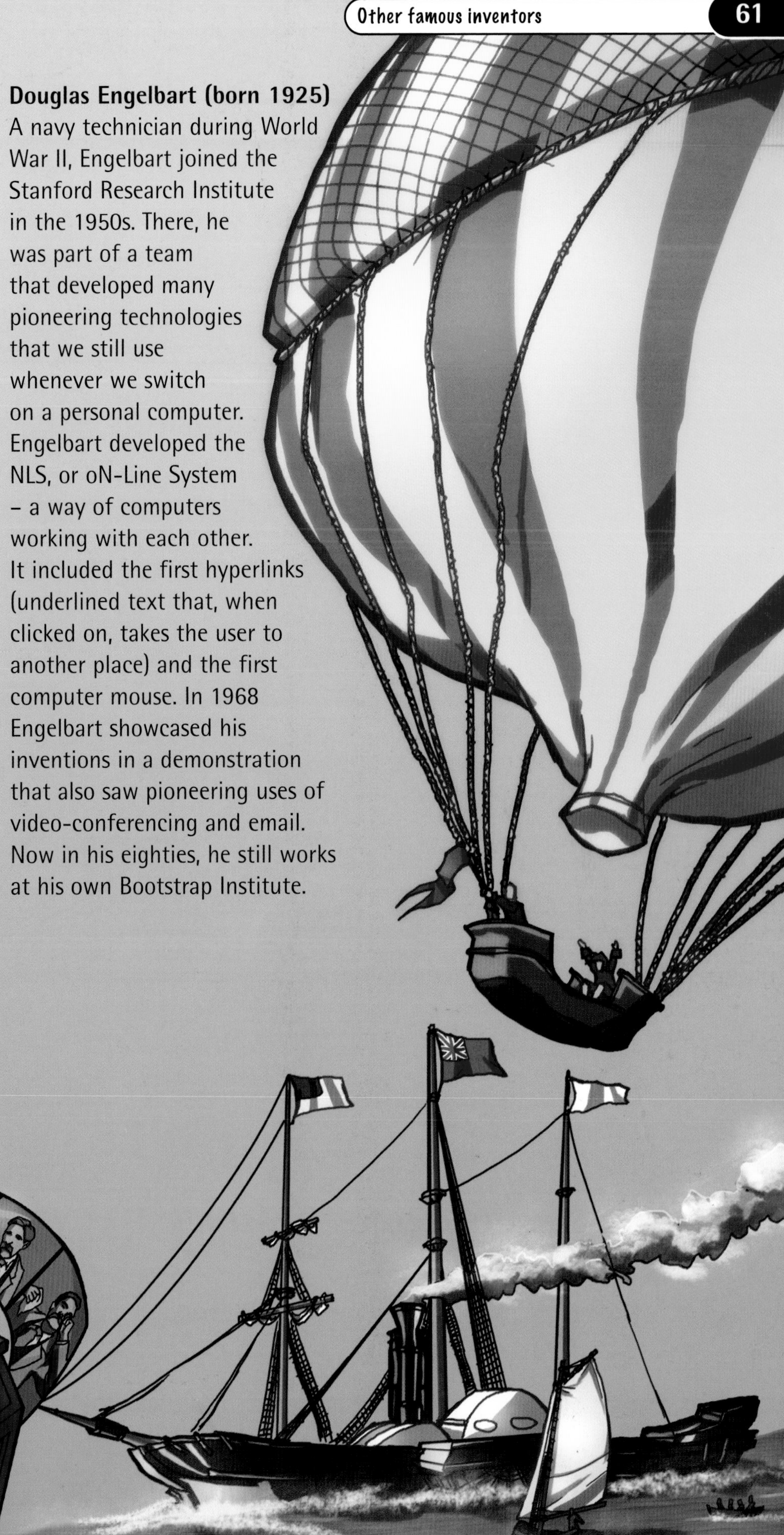

Glossary

Astronomy The study of the Universe and all the stars, planets and other objects in it.

Condense To cool steam so that it turns into liquid water.

Conductor A material or object that can carry an electrical current.

Cylinder In maths, a 3-D shape with a circular cross-section. Tube-shaped cylinders are found inside engines and pumps.

Daylight Saving Time When a country's clocks have moved forward so that afternoons have more daylight and mornings have less.

Density The amount of a substance contained within a certain area.

Dynamo A device that generates electricity.

Electrolysis Passing an electric current through substances in order to separate them.

Element One of over 100 different substances, such as oxygen, gold and hydrogen, that cannot be broken down into simpler substances by chemistry.

Geometry A branch of maths that deals with shapes and sizes.

Glider A light aircraft, with fixed wings but usually no engine, that glides on air currents.

Incandescent Emitting a bright light as a result of being heated.

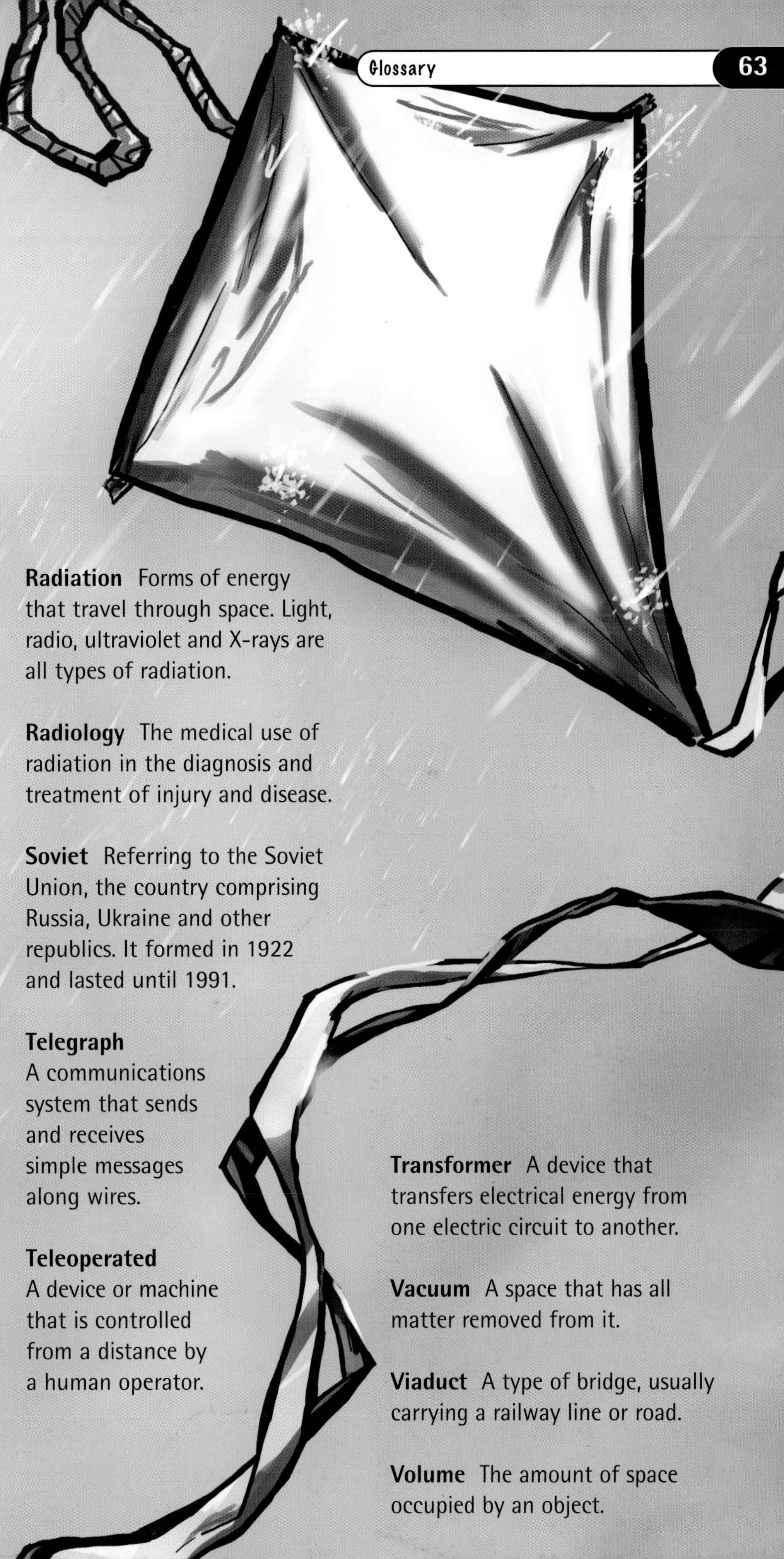

Inquisition An organization of the Roman Catholic Church that prosecuted people accused of holding beliefs opposing the Church.

Magnetism A natural force by which some substances attract or are pulled towards metals like iron.

Nuclear bomb An incredibly powerful and dangerous weapon using the vast energy released from the nuclei of atoms to cause massive damage and loss of life.

Orbit The path of one object around another in space, such as a planet around a star like the Sun or a satellite around Earth.

Pendulum A rod that has a weight hanging from its end, causing it to swing back and forth. Some large clocks use a pendulum.

Physics The branch of science dealing with physical things and the forces that make them behave the way they do.

Pi A number (approximately 3.14159) shown by the symbol π. Pi multiplied by a circle's diameter gives the circle's circumference.

Piston A rod found inside an engine or pump cylinder that moves up and down the cylinder.

Pressure The force of one thing pressing on another.

Pulley A wheel with a rope threaded around it that is used to lift objects.

Radiation Forms of energy that travel through space. Light, radio, ultraviolet and X-rays are all types of radiation.

Radiology The medical use of radiation in the diagnosis and treatment of injury and disease.

Soviet Referring to the Soviet Union, the country comprising Russia, Ukraine and other republics. It formed in 1922 and lasted until 1991.

Telegraph A communications system that sends and receives simple messages along wires.

Teleoperated A device or machine that is controlled from a distance by a human operator.

Transformer A device that transfers electrical energy from one electric circuit to another.

Vacuum A space that has all matter removed from it.

Viaduct A type of bridge, usually carrying a railway line or road.

Volume The amount of space occupied by an object.

Index

aircraft 51, 52–3, 55
aircraft instruments 49
airships 51
Al-Jazari 60
ancient China 4
Anderson, Mary 5
Archimedes 6–9, 11
Archimedes screw 7, 29
Aristotle 11, 13
armonicas 19
astronomy 4, 13, 14, 62

Baldwin, Thomas 50, 51
balloon flight 21, 60
Batchelor, Charles 35, 38, 39
batteries 18, 19, 37, 61
Baylis, Trevor 5
Becquerel, Henri 45
Bell, Alexander Graham 34, 51
bifocal glasses 15, 20
Birdseye, Clarence 61
Biro, Laszlo 5
Boulton, Matthew 21, 24
Brunel, Isambard Kingdom 25, 26–31

cameras 4, 37, 50
cars 36
Catholic Church 10, 13, 14
Claw of Archimedes 8
Clifton suspension bridge 31
clocks 15, 60
computing 5, 42, 61
Copernicus, Nicolas 13
copying machines 21, 25
Crimean War 29, 30
Curie, Marie 43, 44–9
Curie, Pierre 44, 45, 46
Curtiss, Glenn 49, 50–3

Daylight Saving Time 20, 62
density 6, 9, 62
dynamos 36, 40, 48, 62
Dyson, James 5

Edison, Thomas Alva 31, 32–7, 38, 39
electric light 34, 35, 36, 40, 41, 42
electric motors 18, 39, 40
electricity 18, 36, 37, 38–40, 42, 43
electrolysis 45, 62
Engelbart, Douglas 61

films 37
fire fighting 16, 17
Firestone, Harvey 36
flyball governor 25
Ford, Henry 36
Franklin, Benjamin 15, 16–21
Franklin Stove 17
frozen food 61

Gagarin, Yuri 54, 59
Galileo Galilei 9, 10–15
gearing systems 24
geometry 8, 62
gliders 55, 62
Gulf Stream 20

hydro-electric power 41
hydrostatic balance 11

Ibn-al-Haitham 4
intercontinental ballistic missiles (ICBMs) 56

Joliot-Curie, Frédéric 48, 49
Joliot-Curie, Irene 46, 47, 48, 49

kinetoscopes 37
Korolev, Sergei 53, 54–9

lenses 12, 13, 15
lightbulbs 34, 35
lightning rods 18, 19

Marconi, Guglielmo 41, 43
mathematics 7, 8, 11, 13, 26, 44
microphones 34
mining 23, 24, 26
Montgolfier, Jacques and Joseph 60
Moon 13, 57, 59
motorcycles 50, 51

naval aviation 52, 53

odometers 20

patents 17, 24, 33
pendulums 11, 15, 63
phonographs 34
physics 15, 44, 45, 61, 63
pitchblende 44, 45
polonium 45
power stations 36, 41
Priestley, Joseph 21, 25
pulleys 7, 8, 63
pumps 12, 22, 23, 24, 60

radiation 44, 45, 49, 63
radio 5, 41, 42, 43, 52
radioactivity 44, 48
radiology 46, 47, 48, 63
radium 44, 45, 46, 48, 49
radon gas 46
railways 25, 27–8, 30, 31, 32
rockets 54, 55, 56, 57, 59

satellites 56–7
screw propellers 23, 29, 30
seaplanes 52, 53, 54, 55
sectors 12
seismometers 5
spacecraft 54, 56–7, 58–9
steam engines 22, 23, 24, 25
steamships 25, 28–9, 30–1

telegraph 31, 32, 33, 63
telephones 34, 36
telescopes 12, 13
Tesla coil 41, 42
Tesla, Nikola 37, 38–43
tunnels 26–7, 28
Tupolev, Sergei 55, 56

Universal Stock Printer 24
uranium 45

Volta, Alessandro 61
Voltaic Pile 61
voting machines 33

Wardenclyffe Tower 42, 43
Watt, James 21, 22–5
weight 6, 11
World War I 37, 47, 48, 53
Wright, Orville and Wilbur 51

X-rays 43, 47, 48

Zhang Heng 5
Zuse, Konrad 5